Progress in Mathematical Physics
Volume 52

Gravitation and Experiment

Poincaré Seminar 2006

Thibault Damour
Bertrand Duplantier
Vincent Rivasseau
Editors

Birkhäuser Verlag
Basel · Boston · Berlin

Author:

Thibault Damour
Institut des Hautes Études Scientifiques
Route des Chartres
91440 Bures-sur-Yvette
France
e-mail: damour@ihes.fr

Vincent Rivasseau
Laboratoire de Physique Théorique
Université Paris XI
91405 Orsay Cedex
France
e-mail: Vincent.Rivasseau@th.u-psud.fr

Bertrand Duplantier
Service de Physique Théorique
Orme des Merisiers
CEA - Saclay
91191 Gif-sur-Yvette Cedex
France
e-mail: bertrand.duplantier@cea.fr

2000 Mathematics Subject Classification: primary 83E30, 83D05, 83C35, 83C57
secondary 83Cxx, 83-03, 83-05, 85-99, 01A61

Library of Congress Control Number: 2007931898

Bibliographic information published by Die Deutsche Bibliothek
Die Deutsche Bibliothek lists this publication in the Deutsche Nationalbibliografie; detailed bibliographic data is available in the Internet at <http://dnb.ddb.de>.

ISBN 978-3-7643-8523-1 Birkhäuser Verlag, Basel – Boston – Berlin

Printed on acid-free paper produced of chlorine-free pulp. TCF ∞
Printed in Germany

ISBN 987-3-7643-8523-1 e-ISBN: 978-3-7643-8524-8

9 8 7 6 5 4 3 2 1 www.birkhauser.ch

Contents

Foreword

This book is the sixth in a series of lectures of the *Séminaire Poincaré*, which is directed towards a large audience of physicists and of mathematicians.

The goal of this seminar is to provide up-to-date information about general topics of great interest in physics. Both the theoretical and experimental aspects are covered, with some historical background. Inspired by the Bourbaki seminar in mathematics in its organization, hence nicknamed "Bourbaphi", the Poincaré Seminar is held twice a year at the Institut Henri Poincaré in Paris, with contributions prepared in advance. Particular care is devoted to the pedagogical nature of the presentations so as to fulfill the goal of being readable by a large audience of scientists.

This volume contains the ninth such Seminar, held in 2006. It is devoted to Relativity and Experiment.

This book starts with a detailed introduction to general relativity by T. Damour. It includes a review of what may lie beyond by string theorist I. Antoniadis, and collects up-to-date essays on the experimental tests of this theory. General relativity is now a theory well confirmed by detailed experiments, including the precise timing of the double pulsar J0737-3039 explained by M. Kramer, member of the team which discovered it in 2003, and satellite missions such as Gravity Probe B described by J. Mester. The search for detecting gravitational waves is also very much under way as reviewed by J.Y. Vinet.

We hope that the continued publication of this series will serve the community of physicists and mathematicians at professional or graduate student level.

We thank the Commissariat à l'Énergie Atomique (Division des Sciences de la Matière) and the Daniel Iagolnitzer Foundation for sponsoring the Seminar. Special thanks are due to Chantal Delongeas for the preparation of the manuscript.

Thibault Damour
Bertrand Duplantier
Vincent Rivasseau

Séminaire Poincaré
Gravitation & Expérience
Samedi 28 octobre 2006
cat is having a walk in the XXIst century. Eddy Collin, 2004
T. Damour : Einstein aujourd'hui • 10h
I. Antoniadis : Beyond Einstein's Gravity • 11h
J. Mester : Probing Einstein in Space • 14h
M. Kramer : The Double Pulsar • 15h
J.-Y. Vinet : Antennes gravitationnelles • 16h
INSTITUT HENRI POINCARÉ • Amphi Hermite
11, rue Pierre et Marie Curie • 75005 Paris
www.lpthe.jussieu.fr/poincare
ÉCOLE POLYTECHNIQUE
CENTRE NATIONAL DE LA RECHERCHE SCIENTIFIQUE
cea
FONDATION IAGOLNITZER

Gravitation and Experiment, 1–49

Poincaré Seminar 2006

General Relativity Today*

Thibault Damour

Abstract. After recalling the conceptual foundations and the basic structure of general relativity, we review some of its main modern developments (apart from cosmology): (i) the post-Newtonian limit and weak-field tests in the solar system, (ii) strong gravitational fields and black holes, (iii) strong-field and radiative tests in binary pulsar observations, (iv) gravitational waves, (v) general relativity and quantum theory.

1. Introduction

The *theory of general relativity* was developed by Einstein in work that extended from 1907 to 1915. The starting point for Einstein's thinking was the composition of a review article in 1907 on what we today call the *theory of special relativity*. Recall that the latter theory sprang from a new kinematics governing length and time measurements that was proposed by Einstein in June of 1905 [1], [2], following important pioneering work by Lorentz and Poincaré. The theory of special relativity essentially poses a new fundamental framework (in place of the one posed by Galileo, Descartes, and Newton) for the formulation of physical laws: this framework being the chrono-geometric space-time structure of Poincaré and Minkowski. After 1905, it therefore seemed a natural task to formulate, reformulate, or modify the then known physical laws so that they fit within the framework of special relativity. For Newton's law of gravitation, this task was begun (before Einstein had even supplied his conceptual crystallization in 1905) by Lorentz (1900) and Poincaré (1905), and was pursued in the period from 1910 to 1915 by Max Abraham, Gunnar Nordström and Gustav Mie (with these latter researchers developing *scalar* relativistic theories of gravitation).

Meanwhile, in 1907, Einstein became aware that gravitational interactions possessed particular characteristics that suggested the necessity of *generalizing* the framework and structure of the 1905 theory of relativity. After many years of intense intellectual effort, Einstein succeeded in constructing a *generalized theory*

*Translated from the French by Eric Novak.

of relativity (or *general relativity*) that proposed a profound modification of the chrono-geometric structure of the space-time of special relativity. In 1915, in place of a simple, neutral arena, given a priori, independently of all material content, space-time became a physical "field" (identified with the gravitational field). In other words, it was now a dynamical entity, both influencing and influenced by the distribution of mass-energy that it contains.

This radically new conception of the structure of space-time remained for a long while on the margins of the development of physics. Twentieth century physics discovered a great number of new physical laws and phenomena while working with the space-time of special relativity as its fundamental framework, as well as imposing the respect of its symmetries (namely the Lorentz-Poincaré group). On the other hand, the theory of general relativity seemed for a long time to be a theory that was both poorly confirmed by experiment and without connection to the extraordinary progress springing from application of quantum theory (along with special relativity) to high-energy physics. This marginalization of general relativity no longer obtains. Today, general relativity has become one of the essential players in cutting-edge science. Numerous high-precision experimental tests have confirmed, in detail, the pertinence of this theory. General relativity has become the favored tool for the description of the macroscopic universe, covering everything from the big bang to black holes, including the solar system, neutron stars, pulsars, and gravitational waves. Moreover, the search for a consistent description of fundamental physics in its entirety has led to the exploration of theories that unify, within a general quantum framework, the description of matter and all its interactions (including gravity). These theories, which are still under construction and are provisionally known as string theories, contain general relativity in a central way but suggest that the fundamental structure of space-time-matter is even richer than is suggested separately by quantum theory and general relativity.

2. Special Relativity

We begin our exposition of the theory of general relativity by recalling the chrono-geometric structure of space-time in the theory of *special* relativity. The structure of Poincaré-Minkowski space-time is given by a generalization of the Euclidean geometric structure of ordinary space. The latter structure is summarized by the formula $L^2 = (\Delta x)^2 + (\Delta y)^2 + (\Delta z)^2$ (a consequence of the Pythagorean theorem), expressing the square of the distance L between two points in space as a sum of the squares of the differences of the (orthonormal) coordinates x, y, z that label the points. The symmetry group of Euclidean geometry is the group of coordinate transformations $(x, y, z) \rightarrow (x', y', z')$ that leave the quadratic form $L^2 = (\Delta x)^2 + (\Delta y)^2 + (\Delta z)^2$ invariant. (This group is generated by translations, rotations, and "reversals" such as the transformation given by reflection in a mirror, for example: $x' = -x$, $y' = y$, $z' = z$.)

The Poincaré-Minkowski space-time is defined as the ensemble of *events* (idealizations of what happens at a particular point in space, at a particular moment in time), together with the notion of a *(squared) interval* S^2 defined between any two events. An event is fixed by four coordinates, x, y, z, and t, where (x, y, z) are the spatial coordinates of the point in space where the event in question "occurs," and where t fixes the instant when this event "occurs." Another event will be described (within the same reference frame) by four different coordinates, let us say $x + \Delta x$, $y + \Delta y$, $z + \Delta z$, and $t + \Delta t$. The points in space where these two events occur are separated by a distance L given by the formula above, $L^2 = (\Delta x)^2 + (\Delta y)^2 + (\Delta z)^2$. The moments in time when these two events occur are separated by a time interval T given by $T = \Delta t$. The squared interval S^2 between these two events is given as a function of these quantities, by definition, through the following generalization of the Pythagorean theorem:

$$S^2 = L^2 - c^2 T^2 = (\Delta x)^2 + (\Delta y)^2 + (\Delta z)^2 - c^2(\Delta t)^2 , \tag{1}$$

where c denotes the speed of light (or, more precisely, the maximum speed of signal propagation).

Equation (1) defines the *chrono-geometry* of Poincaré-Minkowski space-time. The symmetry group of this chrono-geometry is the group of coordinate transformations $(x, y, z, t) \to (x', y', z', t')$ that leave the quadratic form (1) of the interval S invariant. We will show that this group is made up of linear transformations and that it is generated by translations in space and time, spatial rotations, "boosts" (meaning special Lorentz transformations), and reversals of space and time.

It is useful to replace the time coordinate t by the "light-time" $x^0 \equiv ct$, and to collectively denote the coordinates as $x^\mu \equiv (x^0, x^i)$ where the Greek indices $\mu, \nu, \ldots = 0, 1, 2, 3$, and the Roman indices $i, j, \ldots = 1, 2, 3$ (with $x^1 = x$, $x^2 = y$, and $x^3 = z$). Equation (1) is then written

$$S^2 = \eta_{\mu\nu} \, \Delta x^\mu \, \Delta x^\nu , \tag{2}$$

where we have used the Einstein summation convention[1] and where $\eta_{\mu\nu}$ is a diagonal matrix whose only non-zero elements are $\eta_{00} = -1$ and $\eta_{11} = \eta_{22} = \eta_{33} = +1$. The symmetry group of Poincaré-Minkowski space-time is therefore the ensemble of Lorentz-Poincaré transformations,

$$x'^\mu = \Lambda^\mu_\nu \, x^\nu + a^\mu , \tag{3}$$

where $\eta_{\alpha\beta} \, \Lambda^\alpha_\mu \, \Lambda^\beta_\nu = \eta_{\mu\nu}$.

The chrono-geometry of Poincaré-Minkowski space-time can be visualized by representing, around each point x in space-time, the locus of points that are separated from the point x by a unit (squared) interval, in other words the ensemble of points x' such that $S^2_{xx'} = \eta_{\mu\nu}(x'^\mu - x^\mu)(x'^\nu - x^\nu) = +1$. This locus is a one-sheeted (unit) hyperboloid.

If we were within an ordinary Euclidean space, the ensemble of points x' would trace out a (unit) sphere centered on x, and the "field" of these spheres

[1] Every repeated index is supposed to be summed over all of its possible values.

centered on each point x would allow one to completely characterize the Euclidean geometry of the space. Similarly, in the case of Poincaré-Minkowski space-time, the "field" of unit hyperboloids centered on each point x is a visual characterization of the geometry of this space-time. See Figure 1. This figure gives an idea of the symmetry group of Poincaré-Minkowski space-time, and renders the rigid and homogeneous nature of its geometry particularly clear.

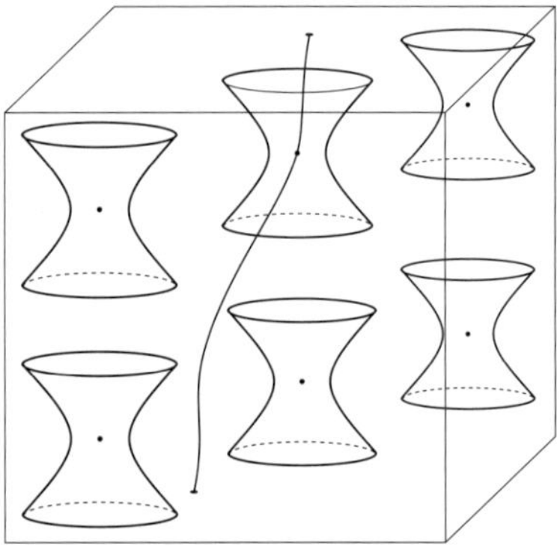

FIGURE 1. Geometry of the "rigid" space-time of the theory of special relativity. This geometry is visualized by representing, around each point x in space-time, the locus of points separated from the point x by a unit (squared) interval. The space-time shown here has only three dimensions: one time dimension (represented vertically), $x^0 = ct$, and two spatial dimensions (represented horizontally), x, y. We have also shown the 'space-time line', or 'world-line', (moving from the bottom to the top of the "space-time block," or from the past towards the future) representing the history of a particle's motion.

The essential idea in Einstein's article of June 1905 was to impose the group of transformations (3) as a symmetry group of the fundamental laws of physics ("the principle of relativity"). This point of view proved to be extraordinarily fruitful, since it led to the discovery of new laws and the prediction of new phenomena. Let us mention some of these for the record: the relativistic dynamics of classical particles, the dilation of lifetimes for relativistic particles, the relation $E = mc^2$ between energy and inertial mass, Dirac's relativistic theory of quantum spin $\frac{1}{2}$ particles, the prediction of antimatter, the classification of particles by rest mass and spin, the relation between spin and statistics, and the CPT theorem.

After these recollections on special relativity, let us discuss the special feature of gravity which, in 1907, suggested to Einstein the need for a profound generalization of the chrono-geometric structure of space-time.

3. The Principle of Equivalence

Einstein's point of departure was a striking experimental fact: all bodies in an external gravitational field fall with the same acceleration. This fact was pointed out by Galileo in 1638. Through a remarkable combination of logical reasoning, thought experiments, and real experiments performed on inclined planes,[2] Galileo was in fact the first to conceive of what we today call the "universality of free-fall" or the "weak principle of equivalence." Let us cite the conclusion that Galileo drew from a hypothetical argument where he varied the ratio between the densities of the freely falling bodies under consideration and the resistance of the medium through which they fall: "Having observed this I came to the conclusion that in a medium totally devoid of resistance all bodies would fall with the same speed" [3]. This universality of free-fall was verified with more precision by Newton's experiments with pendulums, and was incorporated by him into his theory of gravitation (1687) in the form of the identification of the inertial mass m_i (appearing in the fundamental law of dynamics $\boldsymbol{F} = m_i\,\boldsymbol{a}$) with the gravitational mass m_g (appearing in the gravitational force, $F_g = G\,m_g\,m'_g/r^2$):

$$m_i = m_g\,. \tag{4}$$

At the end of the nineteenth century, Baron Roland von Eötvös verified the equivalence (4) between m_i and m_g with a precision on the order of 10^{-9}, and Einstein was aware of this high-precision verification. (At present, the equivalence between m_i and m_g has been verified at the level of 10^{-12} [4].) The point that struck Einstein was that, given the precision with which $m_i = m_g$ was verified, and given the equivalence between inertial mass and energy discovered by Einstein in September of 1905 [2] ($E = m_i\,c^2$), one must conclude that all of the various forms of energy that contribute to the inertial mass of a body (rest mass of the elementary constituents, various binding energies, internal kinetic energy, etc.) do contribute in a strictly identical way to the gravitational mass of this body, meaning both to its capacity for reacting to an external gravitational field and to its capacity to create a gravitational field.

In 1907, Einstein realized that the equivalence between m_i and m_g implicitly contained a deeper equivalence between inertia and gravitation that had important consequences for the notion of an inertial reference frame (which was a fundamental concept in the theory of special relativity). In an ingenious thought experiment, Einstein imagined the behavior of rigid bodies and reference clocks within a freely falling elevator. Because of the universality of free-fall, all of the objects in such a "freely falling local reference frame" would appear not to be accelerating with respect to it. Thus, with respect to such a reference frame, the exterior gravitational field is "erased" (or "effaced"). Einstein therefore postulated what he called the "principle of equivalence" between gravitation and inertia. This principle has two

[2]The experiment with falling bodies said to be performed from atop the Leaning Tower of Pisa is a myth, although it aptly summarizes the essence of Galilean innovation.

parts, that Einstein used in turns. The first part says that, for any external gravitational field whatsoever, it is possible to locally "erase" the gravitational field by using an appropriate freely falling local reference frame and that, because of this, the non-gravitational physical laws apply within this local reference frame just as they would in an inertial reference frame (free of gravity) in special relativity. The second part of Einstein's equivalence principle says that, by starting from an inertial reference frame in special relativity (in the absence of any "true" gravitational field), one can create an apparent gravitational field in a local reference frame, if this reference frame is accelerated (be it in a straight line or through a rotation).

4. Gravitation and Space-Time Chrono-Geometry

Einstein was able (through an extraordinary intellectual journey that lasted eight years) to construct a new theory of gravitation, based on a rich generalization of the 1905 theory of relativity, starting just from the equivalence principle described above. The first step in this journey consisted in understanding that the principle of equivalence would suggest a profound modification of the chrono-geometric structure of Poincaré-Minkowski space-time recalled in Equation (1) above.

To illustrate, let X^α, $\alpha = 0,1,2,3$, be the space-time coordinates in a local, freely-falling reference frame (or *locally inertial reference frame*). In such a reference frame, the laws of special relativity apply. In particular, the infinitesimal space-time interval $ds^2 = dL^2 - c^2\,dT^2$ between two neighboring events within such a reference frame X^α, $X'^\alpha = X^\alpha + dX^\alpha$ (close to the center of this reference frame) takes the form

$$ds^2 = dL^2 - c^2\,dT^2 = \eta_{\alpha\beta}\,dX^\alpha\,dX^\beta\,, \tag{5}$$

where we recall that the repeated indices α and β are summed over all of their values ($\alpha, \beta = 0,1,2,3$). We also know that in special relativity the local energy and momentum densities and fluxes are collected into the ten components of the *energy-momentum tensor* $T^{\alpha\beta}$. (For example, the energy density per unit volume is equal to T^{00}, in the reference frame described by coordinates $X^\alpha = (X^0, X^i)$, $i = 1,2,3$.) The conservation of energy and momentum translates into the equation $\partial_\beta\,T^{\alpha\beta} = 0$, where $\partial_\beta = \partial/\partial\,X^\beta$.

The theory of special relativity tells us that we can change our locally inertial reference frame (while remaining in the neighborhood of a space-time point where one has "erased" gravity) through a Lorentz transformation, $X'^\alpha = \Lambda^\alpha_\beta\,X^\beta$. Under such a transformation, the infinitesimal interval ds^2, Equation (5), remains invariant and the ten components of the (symmetric) tensor $T^{\alpha\beta}$ are transformed according to $T'^{\alpha\beta} = \Lambda^\alpha_\gamma\,\Lambda^\beta_\delta\,T^{\gamma\delta}$. On the other hand, when we pass from a *locally* inertial reference frame (with coordinates X^α) to an *extended* non-inertial reference frame (with coordinates x^μ; $\mu = 0,1,2,3$), the transformation connecting the X^α to the x^μ is no longer a *linear* transformation (like the Lorentz transformation) but becomes a *non-linear* transformation $X^\alpha = X^\alpha(x^\mu)$ that can take any form whatsoever. Because of this, the value of the infinitesimal interval ds^2, when expressed in a general, extended reference frame, will take a more complicated form than the very simple one given by Equation (5) that it had in a reference frame that was locally in free-fall. In fact, by differentiating the non-linear functions $X^\alpha = X^\alpha(x^\mu)$ we obtain the relation $dX^\alpha = \partial X^\alpha/\partial x^\mu \, dx^\mu$. By substituting this relation into (5) we then obtain

$$ds^2 = g_{\mu\nu}(x^\lambda)\, dx^\mu \, dx^\nu \,, \tag{6}$$

where the indices μ, ν are summed over $0, 1, 2, 3$ and where the ten functions $g_{\mu\nu}(x)$ (symmetric over the indices μ and ν) of the four variables x^λ are defined, point by point (meaning that for each point x^λ we consider a reference frame that is locally freely falling at x, with local coordinates X^α_x) by $g_{\mu\nu}(x) = \eta_{\alpha\beta}\, \partial X^\alpha_x(x)/\partial x^\mu \, \partial X^\beta_x(x)/\partial x^\nu$. Because of the nonlinearity of the functions $X^\alpha(x)$, the functions $g_{\mu\nu}(x)$ generally depend in a nontrivial way on the coordinates x^λ.

The local chrono-geometry of space-time thus appears to be given, not by the simple Minkowskian metric (2), with constant coefficients $\eta_{\mu\nu}$, but by a quadratic metric of a much more general type, Equation (6), with coefficients $g_{\mu\nu}(x)$ that vary from point to point. Such general metric spaces had been introduced and studied by Gauss and Riemann in the nineteenth century (in the case where the quadratic form (6) is positive definite). They carry the name *Riemannian spaces* or *curved spaces.* (In the case of interest for Einstein's theory, where the quadratic form (6) is not positive definite, one speaks of a pseudo-Riemannian metric.)

We do not have the space here to explain in detail the various geometric structures in a Riemannian space that are derivable from the data of the infinitesimal interval (6). Let us note simply that given Equation (6), which gives the distance ds between two infinitesimally separated points, we are able, through integration along a curve, to define the length of an arbitrary curve connecting two widely separated points A and B: $L_{AB} = \int_A^B ds$. One can then define the "straightest possible line" between two given points A and B to be the shortest line, in other words the curve that minimizes (or, more generally, extremizes) the integrated distance L_{AB}. These straightest possible lines are called *geodesic curves.* To give a simple example, the geodesics of a spherical surface (like the surface of the Earth) are the great circles (with radius equal to the radius of the sphere). If one mathematically writes the condition for a curve, as given by its parametric representation $x^\mu = x^\mu(s)$, where s is the length along the curve, to extremize the total length L_{AB} one finds that $x^\mu(s)$ must satisfy the following second-order differential equation:

$$\frac{d^2 x^\lambda}{ds^2} + \Gamma^\lambda_{\mu\nu}(x)\, \frac{dx^\mu}{ds} \frac{dx^\nu}{ds} = 0 \,, \tag{7}$$

where the quantities $\Gamma^\lambda_{\mu\nu}$, known as the *Christoffel coefficients* or *connection coefficients*, are calculated, at each point x, from the *metric components* $g_{\mu\nu}(x)$ by the equation

$$\Gamma^\lambda_{\mu\nu} \equiv \frac{1}{2} g^{\lambda\sigma} (\partial_\mu \, g_{\nu\sigma} + \partial_\nu \, g_{\mu\sigma} - \partial_\sigma \, g_{\mu\nu}) \,, \tag{8}$$

where $g^{\mu\nu}$ denotes the matrix inverse to $g_{\mu\nu}$ ($g^{\mu\sigma}\, g_{\sigma\nu} = \delta^\mu_\nu$ where the Kronecker symbol δ^μ_ν is equal to 1 when $\mu = \nu$ and 0 otherwise) and where $\partial_\mu \equiv \partial/\partial x^\mu$ denotes the partial derivative with respect to the coordinate x^μ. To give a very simple example: in the Poincaré-Minkowski space-time the components of the metric are constant, $g_{\mu\nu} = \eta_{\mu\nu}$ (when we use an inertial reference frame). Because of this, the connection coefficients (8) vanish in an inertial reference frame, and the differential equation for geodesics reduces to $d^2 x^\lambda / ds^2 = 0$, whose solutions are ordinary straight lines: $x^\lambda(s) = a^\lambda\, s + b^\lambda$. On the other hand, in a general "curved" space-time (meaning one with components $g_{\mu\nu}$ that depend in an arbitrary way on the point x) the geodesics cannot be *globally* represented by straight lines. One can nevertheless show that it always remains possible, for any $g_{\mu\nu}(x)$ whatsoever, to change coordinates $x^\mu \rightarrow X^\alpha(x)$ in such a way that the connection coefficients $\Gamma^\alpha_{\beta\gamma}$, in the new system of coordinates X^α, vanish *locally*, at a given point X^α_0 (or even along an arbitrary curve). Such *locally geodesic* coordinate systems realize Einstein's equivalence principle mathematically: up to terms of second order, the components $g_{\alpha\beta}(X)$ of a "curved" metric in locally geodesic coordinates X^α ($ds^2 = g_{\alpha\beta}(X)\, dX^\alpha\, dX^\beta$) can be identified with the components of a "flat" Poincaré-Minkowski metric: $g_{\alpha\beta}(X) = \eta_{\alpha\beta} + \mathcal{O}((X - X_0)^2)$, where X_0 is the point around which we expand.

5. Einstein's Equations: Elastic Space-Time

Having postulated that a consistent relativistic theory of the gravitational field should include the consideration of a far-reaching generalization of the Poincaré-Minkowski space-time, Equation (6), Einstein concluded that the same ten functions $\boldsymbol{g}_{\mu\nu}(x)$ should describe both the **g**eometry of space-time as well as **g**ravitation. He therefore got down to the task of finding which equations must be satisfied by the "geometric-gravitational field" $g_{\mu\nu}(x)$. He was guided in this search by three principles. The first was the *principle of general relativity*, which asserts that in the presence of a gravitational field one should be able to write the fundamental laws of physics (including those governing the gravitational field itself) in the same way in any coordinate system whatsoever. The second was that the "source" of the gravitational field should be the energy-momentum tensor $T^{\mu\nu}$. The third was a principle of *correspondence* with earlier physics: in the limit where one neglects gravitational effects, $g_{\mu\nu}(x) = \eta_{\mu\nu}$ should be a solution of the equations being sought, and there should also be a so-called *Newtonian* limit where the new theory reduces to Newton's theory of gravity.

Note that the principle of general relativity (contrary to appearances and contrary to what Einstein believed for several years) has a different physical status than the principle of special relativity. The principle of special relativity was a symmetry principle for the structure of space-time that asserted that physics is *the same* in a particular class of reference frames, and therefore that certain "corresponding" phenomena occur in exactly the same way in different reference frames ("active" transformations). On the other hand, the principle of general relativity is a *principle of indifference*: the phenomena do not (in general) take place in the same way in different coordinate systems. However, none of these (extended) coordinate systems enjoys any privileged status with respect to the others.

The principle asserting that the energy-momentum tensor $T^{\mu\nu}$ should be the source of the gravitational field is founded on two ideas: the relations $E = m_i\, c^2$ and the weak principle of equivalence $m_i = m_g$ show that, in the Newtonian limit, the source of gravitation, the gravitational mass m_g, is equal to the total energy of the body considered, or in other words the integral over space of the energy density T^{00}, up to the factor c^{-2}. Therefore at least one of the components of the tensor $T^{\mu\nu}$ must play the role of source for the gravitational field. However, since the gravitational field is encoded, according to Einstein, by the ten components of the metric $g_{\mu\nu}$, it is natural to suppose that the source for $g_{\mu\nu}$ must also have ten components, which is precisely the case for the (symmetric) tensor $T^{\mu\nu}$.

In November of 1915, after many years of conceptually arduous work, Einstein wrote the final form of the theory of general relativity [6]. *Einstein's equations* are non-linear, second-order partial differential equations for the geometric-gravitational field $g_{\mu\nu}$, containing the energy-momentum tensor $T_{\mu\nu} \equiv g_{\mu\kappa}\, g_{\nu\lambda}\, T^{\kappa\lambda}$ on the right-hand side. They are written as follows:

$$R_{\mu\nu} - \frac{1}{2}\, R\, g_{\mu\nu} = \frac{8\pi\, G}{c^4}\, T_{\mu\nu} \tag{9}$$

where G is the (Newtonian) gravitational constant, c is the speed of light, and $R \equiv g^{\mu\nu}\, R_{\mu\nu}$ and the *Ricci tensor* $R_{\mu\nu}$ are calculated as a function of the connection coefficients $\Gamma^{\lambda}_{\mu\nu}$ (8) in the following way:

$$R_{\mu\nu} \equiv \partial_\alpha\, \Gamma^{\alpha}_{\mu\nu} - \partial_\nu\, \Gamma^{\alpha}_{\mu\alpha} + \Gamma^{\alpha}_{\beta\alpha}\, \Gamma^{\beta}_{\mu\nu} - \Gamma^{\alpha}_{\beta\nu}\, \Gamma^{\beta}_{\mu\alpha}\,. \tag{10}$$

One can show that, in a four-dimensional space-time, the three principles we have described previously uniquely determine Einstein's equations (9). It is nevertheless remarkable that these equations may also be developed from points of view that are completely different from the one taken by Einstein. For example, in the 1960s various authors (in particular Feynman, Weinberg and Deser; see references in [4]) showed that Einstein's equations could be obtained from a purely *dynamical* approach, founded on the consistency of interactions of a long-range spin 2 field, without making any appeal, as Einstein had, to the *geometric* notions coming from mathematical work on Riemannian spaces. Let us also note that if we relax one of the principles described previously (as Einstein did in 1917) we can find a generalization of Equation (9) in which one adds the term $+\Lambda\, g_{\mu\nu}$ to the left-hand side, where Λ is the so-called *cosmological constant*. Such a modification was proposed by Einstein in 1917 in order to be able to write down a globally homogeneous and *stationary* cosmological solution. Einstein rejected this additional term after work by Friedmann (1922) showed the existence of *expanding* cosmological solutions of general relativity and after the observational discovery (by Hubble in 1929) of the expanding motion of galaxies within the universe. However, recent cosmological data have once again made this possibility fashionable, although in the fundamental physics of today one tends to believe that a term of the type $\Lambda\, g_{\mu\nu}$ should be considered as a particular physical contribution to the right-hand side of Einstein's equations (more precisely, as the stress-energy tensor of the *vacuum*, $T^V_{\mu\nu} = -\frac{c^4}{8\pi G}\,\Lambda\, g_{\mu\nu}$), rather than as a universal geometric modification of the left-hand side.

Let us now comment on the physical meaning of Einstein's equations (9). The essential new idea is that the chrono-geometric structure of space-time, Equation (6), in other words the structure that underlies all of the measurements that one could locally make of duration, dT, and of distance, dL, (we recall that, locally, $ds^2 = dL^2 - c^2\, dT^2$) is no longer a rigid structure that is given a priori, once and for all (as was the case for the structure of Poincaré-Minkowski space-time), but instead has become a *field*, a dynamical or *elastic* structure, which is created and/or deformed by the presence of an energy-momentum distribution. See Figure 2, which visualizes the "elastic" geometry of space-time in the theory of general relativity by representing, around each point x, the locus of points (assumed to be infinitesimally close to x) separated from x by a constant (squared) interval: $ds^2 = \varepsilon^2$. As in the case of Poincaré-Minkowski space-time (Figure 1), one arrives at a "field" of hyperboloids. However, this field of hyperboloids no longer has a "rigid" and homogeneous structure.

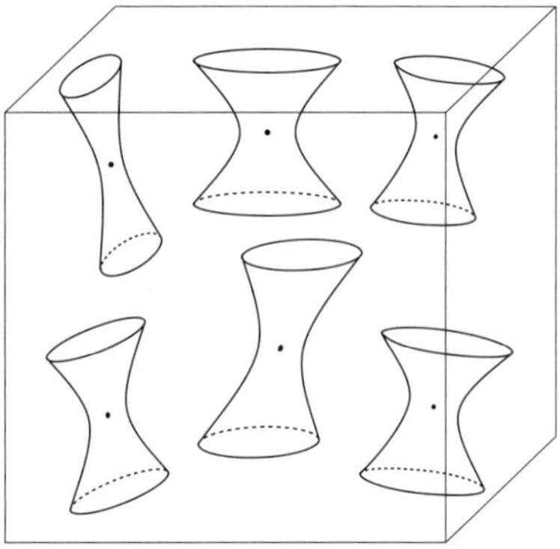

FIGURE 2. "Elastic" space-time geometry in the theory of general relativity. This geometry is visualized by representing, around each space-time point x, the locus of points separated from x by a given small positive (squared) interval.

The *space-time field* $g_{\mu\nu}(x)$ describes the variation from point to point of the chrono-geometry as well as all gravitational effects. The simplest example of space-time chrono-geometric *elasticity* is the effect that the proximity of a mass has on the "local rate of flow for time." In concrete terms, if you separate two twins at birth, with one staying on the surface of the Earth and the other going to live on the peak of a very tall mountain (in other words farther from the Earth's center), and then reunite them after 100 years, the "highlander" will be older (will have lived longer) than the twin who stayed on the valley floor. Everything takes place as if time flows more slowly the closer one is to a given distribution of mass-energy. In mathematical terms this effect is due to the fact that the coefficient $g_{00}(x)$ of $(dx^0)^2$ in Equation (6) is deformed with respect to its value in special relativity, $g_{00}^{\text{Minkowski}} = \eta_{00} = -1$, to become $g_{00}^{\text{Einstein}}(x) \simeq -1 + 2GM/c^2 r$, where M is the Earth's mass (in our example) and r the distance to the center of the Earth. In the example considered above of terrestrial twins the effect is extremely small (a difference in the amount of time lived of about one second over 100 years), but the effect is real and has been verified many times using atomic clocks (see the references in [4]). Let us mention that today this "Einstein effect" has important practical repercussions, for example in aerial or maritime navigation, for the piloting of automobiles, or even farm machinery, etc. In fact, the GPS (Global Positioning System), which uses the data transmitted by a constellation of atomic clocks on board satellites, incorporates the Einsteinian deformation of space-time chrono-geometry into its software. The effect is only on the order of one part in a billion, but if it were not taken into account, it would introduce an unacceptably large error into the GPS, which would continually grow over time. Indeed, GPS performance relies on the high stability of the orbiting atomic clocks, a stability better than 10^{-13}, or in other words 10,000 times greater than the apparent change in frequency($\sim 10^{-9}$) due to the Einsteinian deformation of the chrono-geometry.

6. The Weak-Field Limit and the Newtonian Limit

To understand the physical consequences of Einstein's equations (9), it is useful to begin by considering the limiting case of *weak* geometric-gravitational fields, namely the case where $g_{\mu\nu}(x) = \eta_{\mu\nu} + h_{\mu\nu}(x)$, with perturbations $h_{\mu\nu}(x)$ that are very small with respect to unity: $|h_{\mu\nu}(x)| \ll 1$. In this case, a simple calculation (that we encourage the reader to perform) starting from Definitions (8) and (10) above, leads to the following explicit form of Einstein's equations (where we ignore terms of order h^2 and hT):

$$\Box\, h_{\mu\nu} - \partial_\mu\, \partial^\alpha\, h_{\alpha\nu} - \partial_\nu\, \partial^\alpha\, h_{\alpha\mu} + \partial_{\mu\nu}\, h_\alpha^\alpha = -\frac{16\,\pi\, G}{c^4}\, \tilde{T}_{\mu\nu}\,, \tag{11}$$

where $\Box = \eta^{\mu\nu}\, \partial_{\mu\nu} = \Delta - \partial_0^2 = \partial^2/\partial x^2 + \partial^2/\partial y^2 + \partial^2/\partial z^2 - c^{-2}\, \partial^2/\partial t^2$ denotes the "flat" d'Alembertian (or wave operator; $x^\mu = (ct, x, y, z)$), and where indices in the upper position have been raised by the inverse $\eta^{\mu\nu}$ of the flat metric $\eta_{\mu\nu}$ (numerically $\eta^{\mu\nu} = \eta_{\mu\nu}$, meaning that $-\eta^{00} = \eta^{11} = \eta^{22} = \eta^{33} = +1$). For example $\partial^\alpha\, h_{\alpha\nu}$ denotes $\eta^{\alpha\beta}\, \partial_\alpha\, h_{\beta\nu}$ and $h_\alpha^\alpha \equiv \eta^{\alpha\beta}\, h_{\alpha\beta} = -h_{00} + h_{11} + h_{22} + h_{33}$. The "source" $\tilde{T}_{\mu\nu}$ appearing on the right-hand side of (11) denotes the combination $\tilde{T}_{\mu\nu} \equiv T_{\mu\nu} - \frac{1}{2}\, T_\alpha^\alpha\, \eta_{\mu\nu}$ (when space-time is four-dimensional).

The "linearized" approximation (11) of Einstein's equations is analogous to Maxwell's equations

$$\Box A_\mu - \partial_\mu \partial^\alpha A_\alpha = -4\pi J_\mu , \tag{12}$$

connecting the electromagnetic four-potential $A_\mu \equiv \eta_{\mu\nu} A^\nu$ (where $A^0 = V$, $A^i = \boldsymbol{A}$, $i = 1,2,3$) to the four-current density $J_\mu \equiv \eta_{\mu\nu} J^\nu$ (where $J^0 = \rho$ is the charge density and $J^i = \boldsymbol{J}$ is the current density). Another analogy is that the structure of the left-hand side of Maxwell's equations implies that the "source" J_μ appearing on the right-hand side must satisfy $\partial^\mu J_\mu = 0$ ($\partial^\mu \equiv \eta^{\mu\nu} \partial_\nu$), which expresses the conservation of electric charge. Likewise, the structure of the left-hand side of the linearized form of Einstein's equations (11) implies that the "source" $T_{\mu\nu} = \tilde{T}_{\mu\nu} - \frac{1}{2} \tilde{T}^\alpha_\alpha \eta_{\mu\nu}$ must satisfy $\partial^\mu T_{\mu\nu} = 0$, which expresses the conservation of energy and momentum of matter. (The structure of the left-hand side of the exact form of Einstein's equations (9) implies that the source $T_{\mu\nu}$ must satisfy the more complicated equation $\partial_\mu T^{\mu\nu} + \Gamma^\mu_{\sigma\mu} T^{\sigma\nu} + \Gamma^\nu_{\sigma\mu} T^{\mu\sigma} = 0$, where the terms in ΓT can be interpreted as describing an exchange of energy and momentum between matter and the gravitational field.) The major difference is that, in the case of electromagnetism, the field A_μ and its source J_μ have a single space-time index, while in the gravitational case the field $h_{\mu\nu}$ and its source $\tilde{T}_{\mu\nu}$ have two space-time indices. We shall return later to this analogy/difference between A_μ and $h_{\mu\nu}$ which suggests the existence of a certain relation between gravitation and electromagnetism.

We recover the Newtonian theory of gravitation as the limiting case of Einstein's theory by assuming not only that the gravitational field is a weak deformation of the flat Minkowski space-time ($h_{\mu\nu} \ll 1$), but also that the field $h_{\mu\nu}$ is slowly varying ($\partial_0 h_{\mu\nu} \ll \partial_i h_{\mu\nu}$) and that its source $T_{\mu\nu}$ is non-relativistic ($T_{ij} \ll T_{0i} \ll T_{00}$). Under these conditions Equation (11) leads to a Poisson-type equation for the purely temporal component, h_{00}, of the space-time field,

$$\Delta h_{00} = -\frac{16\pi G}{c^4} \tilde{T}_{00} = -\frac{8\pi G}{c^4} (T_{00} + T_{ii}) \simeq -\frac{8\pi G}{c^4} T_{00} , \tag{13}$$

where $\Delta = \partial_x^2 + \partial_y^2 + \partial_z^2$ is the Laplacian. Recall that, according to Laplace and Poisson, Newton's theory of gravity is summarized by saying that the gravitational field is described by a single potential $U(x)$, produced by the mass density $\rho(x)$ according to the Poisson equation $\Delta U = -4\pi G \rho$, that determines the acceleration of a test particle placed in the exterior field $U(x)$ according to the equation $d^2 x^i / dt^2 = \partial_i U(x) \equiv \partial U / \partial x^i$. Because of the relation $m_i = m_g = E/c^2$ one can identify $\rho = T^{00}/c^2$. We therefore find that (13) reproduces the Poisson equation if $h_{00} = +2U/c^2$. It therefore remains to verify that Einstein's theory indeed predicts that a non-relativistic test particle is accelerated by a space-time field according to $d^2 x^i / dt^2 \simeq \frac{1}{2} c^2 \partial_i h_{00}$. Einstein understood that this was a consequence of the equivalence principle. In fact, as we discussed in Section 4 above, the principle of equivalence states that the gravitational field is (locally) erased in a locally inertial reference frame X^α (such that $g_{\alpha\beta}(X) = \eta_{\alpha\beta} + \mathcal{O}((X - X_0)^2)$). In such a reference frame, the laws of special relativity apply at the point X_0. In particular an isolated (and electrically neutral) body must satisfy a principle of inertia in this frame: its center of mass moves in a straight line at constant speed. In other words it satisfies the equation of motion $d^2 X^\alpha / ds^2 = 0$. By passing back to an arbitrary (extended) coordinate system x^μ, one verifies that this equation for inertial motion transforms into the geodesic equation (7). Therefore (7) describes falling bodies, such as they are observed in arbitrary extended reference frames (for example a reference frame at rest with respect to the Earth or at rest with respect to the center of mass of the solar system). From this one concludes that the relativistic analog of the Newtonian field of gravitational acceleration, $\boldsymbol{g}(x) = \boldsymbol{\nabla} U(x)$, is $g^\lambda(x) \equiv -c^2 \, \Gamma^\lambda_{\mu\nu} \, dx^\mu/ds \, dx^\nu/ds$. By considering a particle whose motion is slow with respect to the speed of light ($dx^i/ds \ll dx^0/ds \simeq 1$) one can easily verify that $g^i(x) \simeq -c^2 \, \Gamma^i_{00}$. Finally, by using the definition (8) of $\Gamma^\alpha_{\mu\nu}$, and the hypothesis of weak fields, one indeed verifies that $g^i(x) \simeq \frac{1}{2} c^2 \, \partial_i \, h_{00}$, in perfect agreement with the identification $h_{00} = 2\,U/c^2$ anticipated above. We encourage the reader to personally verify this result, which contains the very essence of Einstein's theory: gravitational motion is no longer described as being due to a force, but is identified with motion that is "as inertial as possible" within a space-time whose chrono-geometry is deformed in the presence of a mass-energy distribution.

Finding the Newtonian theory as a limiting case of Einstein's theory is obviously a necessity for seriously considering this new theory. But of course, from the very beginning Einstein explored the observational consequences of general relativity that go beyond the Newtonian description of gravitation. We have already mentioned one of these above: the fact that $g_{00} = \eta_{00} + h_{00} \simeq -1 + 2U(x)/c^2$ implies a distortion in the relative measurement of time in the neighborhood of massive bodies. In 1907 (as soon as he had developed the principle of equivalence, and long before he had obtained the field equations of general relativity) Einstein had predicted the existence of such a distortion for measurements of time and frequency in the presence of an external gravitational field. He realized that this should have observable consequences for the frequency, as observed on Earth, of the spectral rays emitted from the surface of the Sun. Specifically, a spectral ray of (proper local) frequency ν_0 emitted from a point x_0 where the (stationary) gravitational potential takes the value $U(\boldsymbol{x}_0)$ and observed (via electromagnetic signals) at a point x where the potential is $U(\boldsymbol{x})$ should appear to have a frequency ν such that

$$\frac{\nu}{\nu_0} = \sqrt{\frac{g_{00}(x_0)}{g_{00}(x)}} \simeq 1 + \frac{1}{c^2} \left[U(\boldsymbol{x}) - U(\boldsymbol{x}_0) \right]. \tag{14}$$

In the case where the point of emission x_0 is in a gravitational potential well deeper than the point of observation x (meaning that $U(\boldsymbol{x}_0) > U(\boldsymbol{x})$) one has $\nu < \nu_0$, in other words a *reddening* effect on frequencies. This effect, which was predicted by Einstein in 1907, was unambiguously verified only in the 1960s, in experiments by Pound and collaborators over a height of about twenty meters. The most precise verification (at the level of $\sim 10^{-4}$) is due to Vessot and collaborators, who compared a hydrogen maser, launched aboard a rocket that reached about

10,000 km in altitude, to a clock of similar construction on the ground. Other experiments compared the times shown on clocks placed aboard airplanes to clocks remaining on the ground. (For references to these experiments see [4].) As we have already mentioned, the "Einstein effect" (14) must be incorporated in a crucial way into the software of satellite positioning systems such as the GPS.

In 1907, Einstein also pointed out that the equivalence principle would suggest that light rays should be deflected by a gravitational field. Indeed, a generalization of the reasoning given above for the motion of particles in an external gravitational field, based on the principle of equivalence, shows that light must itself follow a trajectory that is "as inertial as possible," meaning a geodesic of the curved space-time. Light rays must therefore satisfy the geodesic equation (7). (The only difference from the geodesics followed by material particles is that the parameter s in Equation (7) can no longer be taken equal to the "length" along the geodesic, since a "light" geodesic must also satisfy the constraint $g_{\mu\nu}(x)\,dx^\mu\,dx^\nu = 0$, ensuring that its speed is equal to c, when it is measured in a locally inertial reference frame.) Starting from Equation (7) one can therefore calculate to what extent light is deflected when it passes through the neighborhood of a large mass (such as the Sun). One nevertheless soon realizes that in order to perform this calculation one must know more than the component h_{00} of the gravitational field. The other components of $h_{\mu\nu}$, and in particular the spatial components h_{ij}, come into play in a crucial way in this calculation. This is why it was only in November of 1915, after having obtained the (essentially) final form of his theory, that Einstein could predict the total value of the deflection of light by the Sun. Starting from the linearized form of Einstein's equations (11) and continuing by making the "non-relativistic" simplifications indicated above ($T_{ij} \ll T_{0i} \ll T_{00}$, $\partial_0 h \ll \partial_i h$) it is easy to see that the spatial component h_{ij}, like h_{00}, can be written (after a helpful choice of coordinates) in terms of the Newtonian potential U as $h_{ij}(x) \simeq +2\,U(x)\,\delta_{ij}/c^2$, where δ_{ij} takes the value 1 if $i = j$ and 0 otherwise ($i, j = 1, 2, 3$). By inserting this result, as well as the preceding result $h_{00} = +2\,U/c^2$, into the geodesic equation (7) for the motion of light, one finds (as Einstein did in 1915) that general relativity predicts that the Sun should deflect a ray of light by an angle $\theta = 4GM/(c^2 b)$ where b is the impact parameter of the ray (meaning its minimum distance from the Sun). As is well known, the confirmation of this effect in 1919 (with rather weak precision) made the theory of general relativity and its creator famous.

7. The Post-Newtonian Approximation and Experimental Confirmations in the Regime of Weak and Quasi-Stationary Gravitational Fields

We have already pointed out some of the experimental confirmations of the theory of general relativity. At present, the extreme precision of certain measurements of time or frequency in the solar system necessitates a very careful account of the modifications brought by general relativity to the Newtonian description of spacetime. As a consequence, general relativity is used in a great number of situations, from astronomical or geophysical research (such as very long range radio interferometry, radar tracking of the planets, and laser tracking of the Moon or artificial satellites) to metrological, geodesic or other applications (such as the definition of international atomic time, precision cartography, and the G.P.S.). To do this, the so-called *post-Newtonian* approximation has been developed. This method involves working in the Newtonian limit sketched above while keeping the terms of higher order in the small parameter

$$\varepsilon \sim \frac{v^2}{c^2} \sim |h_{\mu\nu}| \sim |\partial_0 h/\partial_i h|^2 \sim |T^{0i}/T^{00}|^2 \sim |T^{ij}/T^{00}| \,,$$

where v denotes a characteristic speed for the elements in the system considered.

For all present applications of general relativity to the solar system it suffices to include the *first post-Newtonian approximation*, in other words to keep the relative corrections of order ε to the Newtonian predictions. Since the theory of general relativity was poorly verified for a long time, one found it useful (as in the pioneering work of A. Eddington, generalized in the 1960s by K. Nordtvedt and C.M. Will) to study not only the precise predictions of the equations (9) defining Einstein's theory, but to also consider possible deviations from these predictions. These possible deviations were parameterized by means of several non-dimensional "post-Newtonian" parameters. Among these parameters, two play a key role: γ and β. The parameter γ describes a possible deviation from general relativity that comes into play starting at the linearized level, in other words one that modifies the linearized approximation given above. More precisely, it is defined by writing that the difference $h_{ij} \equiv g_{ij} - \delta_{ij}$ between the spatial metric and the Euclidean metric can take the value $h_{ij} = 2\gamma\, U\, \delta_{ij}/c^2$ (in a suitable coordinate system), rather than the value $h_{ij}^{\mathrm{GR}} = 2\, U\, \delta_{ij}/c^2$ that it takes in general relativity, thus differing by a factor γ. Therefore, by definition γ takes the value 1 in general relativity, and $\gamma - 1$ measures the possible deviation with respect to this theory. As for the parameter β (or rather $\beta - 1$), it measures a possible deviation (with respect to general relativity) in the value of $h_{00} \equiv g_{00} - \eta_{00}$. The value of h_{00} in general relativity is $h_{00}^{\mathrm{GR}} = 2\, U/c^2 - 2\, U^2/c^4$, where the first term (discussed above) reproduces the Newtonian approximation (and cannot therefore be modified, as the idea is to parameterize gravitational physics beyond Newtonian predictions) and where the second term is obtained by solving Einstein's equations (9) at the second order of approximation. One then writes an h_{00} of a more general parameterized type, $h_{00} = 2\, U/c^2 - 2\, \beta\, U^2/c^4$, where, by definition, β takes the value 1 in general relativity. Let us finally point out that the parameters $\gamma - 1$ and $\beta - 1$ completely parameterize the post-Newtonian regime of the simplest theoretical alternatives to general relativity, namely the tensor-scalar theories of gravitation. In these theories, the gravitational interaction is carried by two fields at the same time: a massless tensor (spin 2) field coupled to $T^{\mu\nu}$, and a massless scalar (spin 0) field

φ coupled to the trace T^α_α. In this case the parameter $-(\gamma - 1)$ plays the key role of measuring the ratio between the scalar coupling and the tensor coupling.

All of the experiments performed to date within the solar system are compatible with the predictions of general relativity. When they are interpreted in terms of the post-Newtonian (and "post-Einsteinian") parameters $\gamma - 1$ and $\beta - 1$, they lead to strong constraints on possible deviations from Einstein's theory. We make note of the following among tests performed in the solar system: the deflection of electromagnetic waves in the neighborhood of the Sun, the gravitational delay ('Shapiro effect') of radar signals bounced from the Viking lander on Mars, the global analysis of solar system dynamics (including the advance of planetary perihelia), the sub-centimeter measurement of the Earth-Moon distance obtained from laser signals bounced off of reflectors on the Moon's surface, etc. At present (October of 2006) the most precise test (that has been published) of general relativity was obtained in 2003 by measuring the ratio $1 + y \equiv f/f_0$ between the frequency f_0 of radio waves sent from Earth to the Cassini space probe and the frequency f of coherent radio waves sent back (with the same local frequency) from Cassini to Earth and compared (on Earth) to the emitted frequency f_0. The main contribution to the small quantity y is an effect equal, in general relativity, to $y_{GR} = 8(GM/c^3\, b)\, db/dt$ (where b is, as before, the impact parameter) due to the propagation of radio waves in the geometry of a space-time deformed by the Sun: $ds^2 \simeq -(1 - 2\,U/c^2)\, c^2\, dt^2 + (1 + 2\,U/c^2)(dx^2 + dy^2 + dz^2)$, where $U = GM/r$. The maximum value of the frequency change predicted by general relativity was only $|y_{\rm GR}| \lesssim 2 \times 10^{-10}$ for the best observations, but thanks to an excellent frequency stability $\sim 10^{-14}$ (after correction for the perturbations caused by the solar corona) and to a relatively large number of individual measurements spread over 18 days, this experiment was able to verify Einstein's theory at the remarkable level of $\sim 10^{-5}$ [7]. More precisely, when this experiment is interpreted in terms of the post-Newtonian parameters $\gamma - 1$ and $\beta - 1$, it gives the following limit for the parameter $\gamma - 1$ [7]

$$\gamma - 1 = (2.1 \pm 2.3) \times 10^{-5} \,. \tag{15}$$

As for the best present-day limit on the parameter $\beta - 1$, it is smaller than 10^{-3} and comes from the non-observation, in the data from lasers bounced off of the Moon, of any eventual polarization of the Moon's orbit in the direction of the Sun ('Nordtvedt effect'; see [4] for references)

$$4(\beta - 1) - (\gamma - 1) = -0.0007 \pm 0.0010 \,. \tag{16}$$

Although the theory of general relativity is one of the best verified theories in physics, scientists continue to design and plan new or increasingly precise tests of the theory. This is the case in particular for the space mission Gravity Probe B (launched by NASA in April of 2004) whose principal aim is to directly observe a prediction of general relativity that states (intuitively speaking) that space is not only "elastic," but also "fluid." In the nineteenth century Foucault invented both the gyroscope and his famous pendulum in order to render Newton's absolute (and rigid) space directly observable. His experiments in fact showed that, for example, a gyroscope on the surface of the Earth continued, despite the Earth's rotation, to align itself in a direction that is "fixed" with respect to the distant stars. However, in 1918, when Lense and Thirring analyzed some of the consequences of the (linearized) Einstein equations (11), they found that general relativity predicts, among other things, the following phenomenon: the rotation of the Earth (or any other ball of matter) creates a particular deformation of the chrono-geometry of space-time. This deformation is described by the "gravito-magnetic" components h_{0i} of the metric, and induces an effect analogous to the "rotation drag" effect caused by a ball of matter turning in a fluid: the rotation of the Earth (minimally) drags all of the space around it, causing it to continually "turn," as a fluid would.[3] This "rotation of space" translates, in an observable way, into a violation of the effects predicted by Newton and confirmed by Foucault's experiments: in particular, a gyroscope no longer aligns itself in a direction that is "fixed in absolute space," rather its axis of rotation is "dragged" by the rotating motion of the local space where it is located. This effect is much too small to be visible in Foucault's experiments. Its observation by Gravity Probe B (see [8] and the contribution of John Mester to this Poincaré seminar) is important for making Einstein's revolutionary notion of a fluid space-time tangible to the general public.

Up till now we have only discussed the regime of weak and slowly varying gravitational fields. The theory of general relativity predicts the appearance of new phenomena when the gravitational field becomes strong and/or rapidly varying. (We shall not here discuss the cosmological aspects of relativistic gravitation.)

8. Strong Gravitational Fields and Black Holes

The regime of strong gravitational fields is encountered in the physics of *gravitationally condensed bodies.* This term designates the final states of stellar evolution, and in particular neutron stars and black holes. Recall that most of the life of a star is spent slowly burning its nuclear fuel. This process causes the star to be structured as a series of layers of differentiated nuclear structure, surrounding a progressively denser core (an "onion-like" structure). When the initial mass of the star is sufficiently large, this process ends into a catastrophic phenomenon: the core, already much denser than ordinary matter, collapses in on itself under the influence of its gravitational self-attraction. (This implosion of the central part of the star is, in many cases, accompanied by an explosion of the outer layers of the star—a supernova.) Depending on the quantity of mass that collapses with the core of a star, this collapse can give rise to either a neutron star or a black hole.

A *neutron star* condenses a mass on the order of the mass of the Sun inside a radius on the order of 10 km. The density in the interior of a neutron star (named

[3]Recent historical work (by Herbert Pfister) has in fact shown that this effect had already been derived by Einstein within the framework of the provisory relativistic theory of gravity that he started to develop in 1912 in collaboration with Marcel Grossmann.

thus because neutrons dominate its nuclear composition) is more than 100 million tons per cubic centimeter (10^{14} g/cm^3)! It is about the same as the density in the interior of atomic nuclei. What is important for our discussion is that the deformation away from the Minkowski metric in the immediate neighborhood of a neutron star, measured by $h_{00} \sim h_{ii} \sim 2GM/c^2R$, where R is the radius of the star, is no longer a small quantity, as it was in the solar system. In fact, while $h \sim 2GM/c^2R$ is on the order of 10^{-9} for the Earth and 10^{-6} for the Sun, one finds that $h \sim 0.4$ for a typical neutron star ($M \simeq 1.4\,M_\odot$, $R \sim 10$ km). One thus concludes that it is no longer possible, as was the case in the solar system, to study the structure and physics of neutron stars by using the post-Newtonian approximation outlined above. One must consider the exact form of Einstein's equations (9), with all of their non-linear structure. Because of this, we expect that observations concerning neutron stars will allow us to confirm (or refute) the theory of general relativity in its strongly non-linear regime. We shall discuss such tests below in relation to observations of binary pulsars.

A *black hole* is the result of a *continued* collapse, meaning that it does not stop with the formation of an ultra-dense star (such as a neutron star). (The physical concept of a black hole was introduced by J.R. Oppenheimer and H. Snyder in 1939. The global geometric structure of black holes was not understood until some years later, thanks notably to the work of R. Penrose. For a historical review of the idea of black holes see [9].) It is a particular structure of curved space-time characterized by the existence of a boundary (called the "black hole surface" or "horizon") between an exterior region, from which it is possible to emit signals to infinity, and an interior region (of space-time), within which any emitted signal remains trapped. See Figure 3.

The cones shown in this figure are called "light cones." They are defined as the locus of points (infinitesimally close to x) such that $ds^2 = 0$, with $dx^0 = cdt \geq 0$. Each represents the beginning of the space-time history of a flash of light emitted from a certain point in space-time. The cones whose vertices are located outside of the horizon (the shaded zone) will evolve by spreading out to infinity, thus representing the possibility for electromagnetic signals to reach infinity.

On the other hand, the cones whose vertices are located inside the horizon (the grey zone) will evolve without ever succeeding in escaping the grey zone. It is therefore impossible to emit an electromagnetic signal that reaches infinity from the grey zone. The horizon, namely the boundary between the shaded zone and the unshaded zone, is itself the history of a particular flash of light, emitted from the center of the star over the course of its collapse, such that it asymptotically stabilizes as a space-time cylinder. This space-time cylinder (the asymptotic horizon) therefore represents the space-time history of a bubble of light that, viewed locally, moves outward at the speed c, but which globally "runs in place." This remarkable behavior is a striking illustration of the "fluid" character of space-time in Einstein's theory. Indeed, one can compare the preceding situation with what may take place around the open drain of an emptying sink: a wave may move along

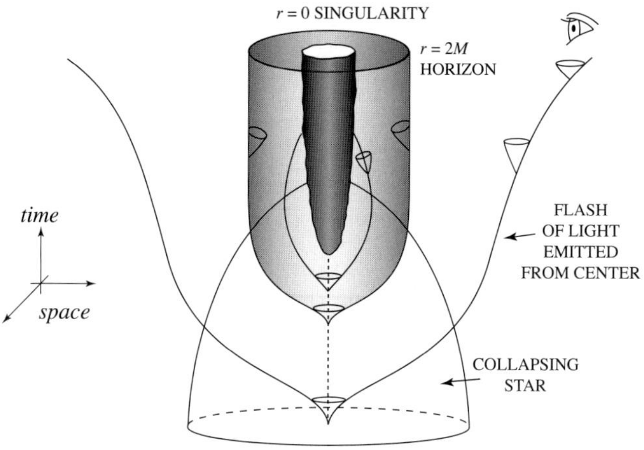

FIGURE 3. Schematic representation of the space-time for a black hole created from the collapse of a spherical star. Each cone represents the space-time history of a flash of light emitted from a point at a particular instant. (Such a "cone field" is obtained by taking the limit $\varepsilon^2 = 0$ from Figure 2, and keeping only the upper part, in other words the part directed towards the future, of the double cones obtained as limits of the hyperboloids of Figure 2.) The interior of the black hole is shaded, its outer boundary being the "black hole surface" or "horizon." The "inner boundary" (shown in dark grey) of the interior region of the black hole is a space-time singularity of the big-crunch type.

the water, away from the hole, all the while running in place with respect to the sink because of the falling motion of the water in the direction of the drain.

Note that the temporal development of the interior region is limited, terminating in a *singularity* (the dark grey surface) where the curvature becomes infinite and where the classical description of space and time loses its meaning. This singularity is locally similar to the temporal inverse of a cosmological singularity of the big bang type. This is called a *big crunch.* It is a space-time frontier, beyond which space-time ceases to exist. The appearance of singularities associated with regions of strong gravitational fields is a generic phenomenon in general relativity, as shown by theorems of R. Penrose and S.W. Hawking.

Black holes have some remarkable properties. First, a *uniqueness* theorem (due to W. Israel, B. Carter, D.C. Robinson, G. Bunting, and P.O. Mazur) asserts that an isolated, stationary black hole (in Einstein-Maxwell theory) is completely described by three parameters: its mass M, its angular momentum J, and its electric charge Q. The exact solution (called the Kerr-Newman solution) of Einstein's equations (11) describing a black hole with parameters M, J, Q is explicitly known. We shall here content ourselves with writing the space-time geometry in the simplest case of a black hole: the one in which $J = Q = 0$ and the black hole is described only by its mass (a solution discovered by K. Schwarzschild in January of 1916):

$$ds^2 = -\left(1 - \frac{2GM}{c^2 r}\right) c^2\,dt^2 + \frac{dr^2}{1 - \frac{2GM}{c^2 r}} + r^2(d\theta^2 + \sin^2\theta\,d\varphi^2)\,. \qquad (17)$$

We see that the purely temporal component of the metric, $g_{00} = -(1 - 2GM/c^2 r)$, vanishes when the radial coordinate r takes the value $r = r_H \equiv 2GM/c^2$. According to the earlier equation (14), it would therefore seem that light emitted from an arbitrary point on the sphere $r_0 = r_H$, when it is viewed by an observer located anywhere in the exterior (in $r > r_H$), would experience an infinite reddening of its emission frequency ($\nu/\nu_0 = 0$). In fact, the sphere $r_H = 2GM/c^2$ is the *horizon* of the Schwarzschild black hole, and no particle (that is capable of emitting light) can remain at rest when $r = r_H$ (nor, a fortiori, when $r < r_H$). To study what happens at the horizon ($r = r_H$) or in the interior ($r < r_H$) of a Schwarzschild black hole, one must use other space-time coordinates than the coordinates (t, r, θ, φ) used in Equation (17). The "big crunch" singularity in the interior of a Schwarzschild black hole, in the coordinates of (17), is located at $r = 0$ (which does not describe, as one might believe, a point in space, but rather an instant in time).

The space-time metric of a black hole space-time, such as Equation (17) in the simple case $J = Q = 0$, allows one to study the influence of a black hole on particles and fields in its neighborhood. One finds that a black hole is a gravitational potential well that is so deep that any particle or wave that penetrates the interior of the black hole (the region $r < r_H$) will never be able to come out again, and that the total energy of the particle or wave that "falls" into the black hole ends up augmenting the total mass-energy M of the black hole. By studying such black hole "accretion" processes with falling particles (following R. Penrose), D. Christodoulou and R. Ruffini showed that a black hole is not only a potential well, but also a physical object possessing a significant *free energy* that it is possible, in principle, to extract. Such *black hole energetics* is encapsulated in the "mass formula" of Christodoulou and Ruffini (in units where $c = 1$)

$$M^2 = \left(M_{\rm irr} + \frac{Q^2}{4\,GM_{\rm irr}}\right)^2 + \frac{J^2}{4\,G^2 M_{\rm irr}^2}\,, \qquad (18)$$

where $M_{\rm irr}$ denotes the *irreducible mass* of the black hole, a quantity *that can only grow, irreversibly.* One deduces from (18) that a rotating ($J \neq 0$) and/or charged ($Q \neq 0$) black hole possesses a free energy $M - M_{\rm irr} > 0$ that can, in principle, be extracted through processes that reduce its angular momentum and/or its electric charge. Such black hole energy-extraction processes may lie at the origin of certain ultra-energetic astrophysical phenomena (such as quasars or gamma ray bursts). Let us note that, according to Equation (18), (rotating or charged) black holes are the largest reservoirs of free energy in the Universe: in fact, 29% of their mass energy can be stored in the form of rotational energy, and up to 50% can be stored in the form of electric energy. These percentages are much higher than the few percent of nuclear binding energy that is at the origin of all the light emitted by stars over their lifetimes. Even though there is not, at present, irrefutable proof of the existence of black holes in the universe, an entire range of very strong presumptive evidence lends credence to their existence. In particular, more than a dozen X-ray emitting binary systems in our galaxy are most likely made up of a black hole and an ordinary star. Moreover, the center of our galaxy seems to contain a very compact concentration of mass $\sim 3 \times 10^6 M_\odot$ that is probably a black hole. (For a review of the observational data leading to these conclusions see, for example, Section 7.6 of the recent book by N. Straumann [6].)

The fact that a quantity associated with a black hole, here the irreducible mass $M_{\rm irr}$, or, according to a more general result due to S.W. Hawking, the total area A of the surface of a black hole ($A = 16\,\pi\, G^2 M_{\rm irr}^2$), can evolve only by irreversibly growing is reminiscent of the second law of thermodynamics. This result led J.D. Bekenstein to interpret the horizon area, A, as being proportional to the *entropy* of the black hole. Such a thermodynamic interpretation is reinforced by the study of the growth of A under the influence of external perturbations, a growth that one can in fact attribute to some local dissipative properties of the black hole surface, notably a surface viscosity and an electrical resistivity equal to 377 ohm (as shown in work by T. Damour and R.L. Znajek). These "thermodynamic" interpretations of black hole properties are based on simple analogies at the level of classical physics, but a remarkable result by Hawking showed that they have real content at the level of quantum physics. In 1974, Hawking discovered that the presence of a horizon in a black hole space-time affected the definition of a quantum particle, and caused a black hole to continuously emit a flux of particles having the characteristic spectrum (Planck spectrum) of thermal emission at the temperature $T = 4\,\hbar\, G\, \partial M/\partial A$, where $\hbar$ is the reduced Planck constant. By using the general thermodynamic relation connecting the temperature to the energy $E = M$ and the entropy S, $T = \partial M/\partial S$, we see from Hawking's result (in conformity with Bekenstein's ideas) that a black hole possesses an *entropy* S equal (again with $c = 1$) to

$$S = \frac{1}{4}\,\frac{A}{\hbar\, G}\,. \tag{19}$$

The Bekenstein-Hawking formula (19) suggests an unexpected, and perhaps profound, connection between gravitation, thermodynamics, and quantum theory. See Section 11 below.

9. Binary Pulsars and Experimental Confirmations in the Regime of Strong and Radiating Gravitational Fields

Binary pulsars are binary systems made up of a pulsar (a rapidly spinning neutron star) and a very dense companion star (either a neutron star or a white dwarf). The first system of this type (called PSR B1913+16) was discovered by R.A. Hulse and J.H. Taylor in 1974 [10]. Today, a dozen are known. Some of these (including the first-discovered PSR B1913+16) have revealed themselves to be remarkable probes of relativistic gravitation and, in particular, of the regime of strong and/or radiating gravitational fields. The reason for which a binary pulsar allows for the probing of strong gravitational fields is that, as we have already indicated above, the deformation of the space-time geometry in the neighborhood of a neutron star is no longer a small quantity, as it is in the solar system. Rather, it is on the order of unity: $h_{\mu\nu} \equiv g_{\mu\nu} - \eta_{\mu\nu} \sim 2GM/c^2R \sim 0.4$. (We note that this value is only 2.5 times smaller than in the extreme case of a black hole, for which $2GM/c^2R = 1$.) Moreover, the fact that the gravitational interaction propagates at the speed of light (as indicated by the presence of the wave operator, $\Box = \Delta - c^{-2}\partial^2/\partial t^2$ in (11)) between the pulsar and its companion is found to play an observationally significant role for certain binary pulsars.

Let us outline how the observational data from binary pulsars are used to probe the regime of strong ($h_{\mu\nu}$ on the order of unity) and/or radiative (effects propagating at the speed c) gravitational fields. (For more details on the observational data from binary pulsars and their use in probing relativistic gravitation, see Michael Kramer's contribution to this Poincaré seminar.) Essentially, a pulsar plays the role of an extremely stable *clock*. Indeed, the "pulsar phenomenon" is due to the rotation of a bundle of electromagnetic waves, created in the neighborhood of the two magnetic poles of a strongly magnetized neutron star (with a magnetic field on the order of 10^{12} Gauss, 10^{12} times the size of the terrestrial magnetic field). Since the magnetic axis of a pulsar is not aligned with its axis of rotation, the rapid rotation of the pulsar causes the (inner) magnetosphere as a whole to rotate, and likewise the bundle of electromagnetic waves created near the magnetic poles. The pulsar is therefore analogous to a lighthouse that sweeps out space with two bundles (one per pole) of electromagnetic waves. Just as for a lighthouse, one does not see the pulsar from Earth except when the bundle sweeps the Earth, thus causing a flash of electromagnetic noise with each turn of the pulsar around itself (in some cases, one even sees a secondary flash, due to emission from the second pole, after each half-turn). One can then measure the time of arrival at Earth of (the center of) each flash of electromagnetic noise. The basic observational data of a pulsar are thus made up of a regular, discrete sequence of the *arrival times* at Earth of these flashes or "pulses." This sequence is analogous to the signal from a clock: tick, tick, tick, ... Observationally, one finds that some pulsars (and in particular those that belong to binary systems) thus define clocks of a stability comparable to the best atomic clocks [11]. In the case of a solitary pulsar, the sequence of its arrival times is (in essence) a regular "arithmetic sequence," $T_N = aN + b$, where N is an integer labeling the pulse considered, and where a is equal to the period of rotation of the pulsar around itself. In the case of a binary pulsar, the sequence of arrival times is a much richer signal, say $T_N = aN + b + \Delta_N$, where Δ_N measures the deviation with respect to a regular arithmetic sequence. This deviation (after the subtraction of effects not connected to the orbital period of the pulsar) is due to a whole ensemble of physical effects connected to the orbital motion of the pulsar around its companion or, more precisely, around the center of mass of the binary system. Some of these effects could be predicted by a purely *Keplerian* description of the motion of the pulsar in space, and are analogous to the "Rœmer effect" that allowed Rœmer to determine, for the first time, the speed of light from the arrival times at Earth of light signals coming from Jupiter's satellites (the light signals coming from a body moving in orbit are "delayed" by the time taken by light to cross this orbit and arrive at Earth). Other effects can only be predicted and calculated by using a *relativistic* description, either of the orbital motion of the pulsar, or of the propagation of electromagnetic signals between the pulsar and Earth. For example, the following facts must be accounted for: (i) the "pulsar clock" moves at a large speed (on the order of 300 km/s $\sim 10^{-3}c$) and is embedded in the varying gravitational potential of the companion; (ii) the orbit of the pulsar is not a simple Keplerian ellipse, but (in general relativity) a more complicated orbit that traces out a "rosette" around the center of mass; (iii) the propagation of electromagnetic signals between the pulsar and Earth takes place in a space-time that is curved by both the pulsar and its companion, which leads to particular effects of relativistic delay; etc. Taking relativistic effects in the theoretical description of arrival times for signals emitted by binary pulsars into account thus leads one to write what is called a *timing formula.* This timing formula (due to T. Damour and N. Deruelle) in essence allows one to parameterize the sequence of arrival times, $T_N = aN + b + \Delta_N$, in other words to parameterize Δ_N, as a function of a set of "phenomenological parameters" that include not only the so-called "Keplerian" parameters (such as the orbital period P, the projection of the semi-major axis of the pulsar's orbit along the line of sight $x_A = a_A \sin i$, and the eccentricity e), but also the *post-Keplerian* parameters associated with the relativistic effects mentioned above. For example, effect (i) discussed above is parameterized by a quantity denoted γ_T; effect (ii) by (among others) the quantities $\dot{\omega}$, $\dot{P}$; effect (iii) by the quantities r, s; etc.

The way in which observations of binary pulsars allow one to test relativistic theories of gravity is therefore the following. A (least-squares) fit between the observational timing data, $\Delta_N^{\rm obs}$, and the parameterized theoretical timing formula, $\Delta_N^{\rm th}(P, x_A, e; \gamma_T, \dot{\omega}, \dot{P}, r, s)$, allows for the determination of the observational values of the Keplerian ($P^{\rm obs}, x_A^{\rm obs}, e^{\rm obs}$) and post-Keplerian ($\gamma_T^{\rm obs}, \dot{\omega}^{\rm obs}, \dot{P}^{\rm obs}, r^{\rm obs}, s^{\rm obs}$) parameters. The theory of general relativity predicts the value of each post-Keplerian parameter as a function of the Keplerian parameters and the two masses of the binary system (the mass m_A of the pulsar and the mass m_B of the companion). For example, the theoretical value predicted by general relativity for the parameter γ_T is $\gamma_T^{\rm GR}(m_A, m_B) = e n^{-1}(GMn/c^3)^{2/3}\, m_B(m_A + 2\, m_B)/M^2$, where e is the eccentricity, $n = 2\pi/P$ the orbital frequency, and $M \equiv m_A + m_B$. We thus see that, if one assumes that general relativity is correct, the observational measurement of a post-Keplerian parameter, for example $\gamma_T^{\rm obs}$, determines a *curve* in the plane (m_A, m_B) of the two masses: $\gamma_T^{\rm GR}(m_A, m_B) = \gamma_T^{\rm obs}$, in our example. The measurement of two post-Keplerian parameters thus gives two curves in the

(m_A, m_B) plane and generically allows one to determine the values of the two masses m_A and m_B, by considering the intersection of the two curves. We obtain a test of general relativity as soon as one observationally measures three or more post-Keplerian parameters: if the three (or more) curves all intersect at one point in the plane of the two masses, the theory of general relativity is confirmed, but if this is not the case the theory is refuted. At present, four distinct binary pulsars have allowed one to test general relativity. These four "relativistic" binary pulsars are: the first binary pulsar PSR B1913+16, the pulsar PSR B1534+12 (discovered by A. Wolszczan in 1991), and two recently discovered pulsars: PSR J1141−6545 (discovered in 1999 by V.M. Kaspi et al., whose first timing results are due to M. Bailes et al. in 2003), and PSR J0737−3039 (discovered in 2003 by M. Burgay et al., whose first timing results are due to A.G. Lyne et al. and M. Kramer et al.). With the exception of PSR J1141−6545, whose companion is a white dwarf, the companions of the pulsars are neutron stars. In the case of PSR J0737−3039 the companion turns out to also be a pulsar that is visible from Earth.

In the system PSR B1913+16, *three* post-Keplerian parameters have been measured $(\dot{\omega}, \gamma_T, \dot{P})$, which gives *one* test of the theory. In the system PSR J1141 -65, *three* post-Keplerian parameters have been measured $(\dot{\omega}, \gamma_T, \dot{P})$, which gives *one* test of the theory. (The parameter s is also measured through scintillation phenomena, but the use of this measurement for testing gravitation is more problematic.) In the system PSR B1534+12, *five* post-Keplerian parameters have been measured, which gives *three* tests of the theory. In the system PSR J0737−3039, *six* post-Keplerian parameters,[4] which gives *four* tests of the theory. It is remarkable that all of these tests have confirmed general relativity. See Figure 4 and, for references and details, [4, 11, 12, 13], as well as the contribution by Michael Kramer.

Note that, in Figure 4, some post-Keplerian parameters are measured with such great precision that they in fact define very thin curves in the m_A, m_B plane. On the other hand, some of them are only measured with a rough fractional precision and thus define "thick curves," or "strips" in the plane of the masses (see, for example, the strips associated with $\dot{P}$, r and s in the case of PSR B1534+12). In any case, the theory is confirmed when all of the strips (thick or thin) have a non-empty common intersection. (One should also note that the strips represented in Figure 4 only use the "one sigma" error bars, in other words a 68% level of confidence. Therefore, the fact that the $\dot{P}$ strip for PSR B1534+12 is a little bit disjoint from the intersection of the other strips is not significant: a "two sigma" figure would show excellent agreement between observation and general relativity.)

In view of the arguments presented above, all of the tests shown in Figure 4 confirm the validity of general relativity in the regime of strong gravitational fields $(h_{\mu\nu} \sim 1)$. Moreover, the four tests that use measurements of the parameter $\dot{P}$ (in

[4]In the case of PSR J0737−3039, one of the six measured parameters is the ratio x_A/x_B between a Keplerian parameter of the pulsar and its analog for the companion, which turns out to also be a pulsar.

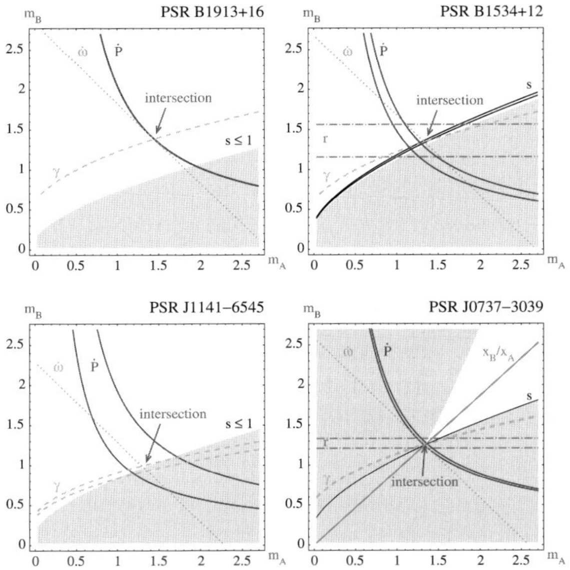

FIGURE 4. Tests of general relativity obtained from observations of four binary pulsars. For each binary pulsar one has traced the "curves," in the plane of the two masses (m_A = mass of the pulsar, m_B = mass of the companion), defined by equating the theoretical expressions for the various post-Keplerian parameters, as predicted by general relativity, to their observational value, determined through a least-squares fit to the parameterized theoretical timing formula. Each "curve" is in fact a "strip," whose thickness is given by the (one sigma) precision with which the corresponding post-Keplerian parameter is measured. For some parameters, these strips are too thin to be visible. The grey zones would correspond to a sine for the angle of inclination of the orbital plane with respect to the plane of the sky that is greater than 1, and are therefore physically excluded.

the four corresponding systems) are direct experimental confirmations of the fact that the gravitational interaction propagates at the speed c between the companion and the pulsar. In fact, $\dot{P}$ denotes the long-term variation $\langle dP/dt \rangle$ of the orbital period. Detailed theoretical calculations of the motion of two gravitationally condensed objects in general relativity, that take into account the effects connected to the propagation of the gravitational interaction at finite speed [14], have shown that one of the observable effects of this propagation is a long-term decrease in the orbital period given by the formula

$$\dot{P}^{\rm GR}(m_A, m_B) = -\frac{192\,\pi}{5} \, \frac{1 + \frac{73}{24}\,e^2 + \frac{37}{96}\,e^4}{(1-e^2)^{7/2}} \left(\frac{GMn}{c^3}\right)^{5/3} \frac{m_A\,m_B}{M^2}\,.$$

The direct physical origin of this decrease in the orbital period lies in the modification, produced by general relativity, of the usual Newtonian law of gravitational attraction between two bodies, $F_{\rm Newton} = G\,m_A\,m_B/r_{AB}^2$. In place of such a simple law, general relativity predicts a more complicated force law that can be expanded in the symbolic form

$$F_{\rm Einstein} = \frac{G\,m_A\,m_B}{r_{AB}^2}\left(1 + \frac{v^2}{c^2} + \frac{v^4}{c^4} + \frac{v^5}{c^5} + \frac{v^6}{c^6} + \frac{v^7}{c^7} + \cdots\right), \tag{20}$$

where, for example, "v^2/c^2" represents a whole set of terms of order v_A^2/c^2, v_B^2/c^2, $v_A\,v_B/c^2$, or even $G\,m_A/c^2\,r$ or $G\,m_B/c^2\,r$. Here v_A denotes the speed of body A, v_B that of body B, and r_{AB} the distance between the two bodies. The term of order v^5/c^5 in Equation (20) is particularly important. This term is a direct consequence of the finite-speed propagation of the gravitational interaction between A and B, and its calculation shows that it contains a component that is opposed to the relative speed $\boldsymbol{v}_A - \boldsymbol{v}_B$ of the two bodies and that, consequently, slows down the orbital motion of each body, causing it to evolve towards an orbit that lies closer to its companion (and therefore has a shorter orbital period). This "braking" term (which is correlated with the emission of gravitational waves), $\delta F_{\rm Einstein} \sim v^5/c^5\,F_{\rm Newton}$, leads to a long-term decrease in the orbital period $\dot{P}^{\rm GR} \sim -(v/c)^5 \sim -10^{-12}$ that is very small, but whose reality has been verified with a fractional precision of order 10^{-3} in PSR B1913+16 and of order 20% in PSR B1534+12 and PSR J1141−6545 [4, 11, 13].

To conclude this brief outline of the tests of relativistic gravitation by binary pulsars, let us note that there is an analog, for the regime of strong gravitational fields, of the formalism of parametrization for possible deviations from general relativity mentioned in Section 6 in the framework of weak gravitational fields (using the post-Newtonian parameters $\gamma - 1$ and $\beta - 1$). This analog is obtained by considering a two-parameter family of relativistic theories of gravitation, assuming that the gravitational interaction is propagated not only by a tensor field $g_{\mu\nu}$ but also by a scalar field φ. Such a class of tensor-scalar theories of gravitation allows for a description of possible deviations in both the solar system and in binary pulsars. It also allows one to explicitly demonstrate that binary pulsars indeed test the effects of strong fields that go beyond the tests of the weak fields of the solar system by exhibiting classes of theories that are compatible with all of the observations in the solar system but that are incompatible with the observations of binary pulsars, see [4, 13].

10. Gravitational Waves: Propagation, Generation, and Detection

As soon as he had finished constructing the theory of general relativity, Einstein realized that it implied the existence of waves of geometric deformations of space-time, or "gravitational waves" [15, 2]. Mathematically, these waves are analogs (with the replacement $A_\mu \rightarrow h_{\mu\nu}$) of electromagnetic waves, but conceptually they signify something remarkable: they exemplify, in the purest possible way, the "elastic" nature of space-time in general relativity. Before Einstein space-time was a rigid structure, given a priori, which was not influenced by the material content of the Universe. After Einstein, a distribution of matter (or more generally of mass-energy) that changes over the course of time, let us say for concreteness a binary system of two neutron stars or two black holes, will not only deform the chrono-geometry of the space-time in its immediate neighborhood, but this deformation will propagate in every possible direction away from the system considered, and will travel out to infinity in the form of a wave whose oscillations will reflect the temporal variations of the matter distribution. We therefore see that the study of these gravitational waves poses three separate problems: that of generation, that of propagation, and, finally, that of detection of such gravitational radiation. These three problems are at present being actively studied, since it is hoped that we will soon detect gravitational waves, and thus will be able to obtain new information about the Universe [16]. We shall here content ourselves with an elementary introduction to this field of research. For a more detailed introduction to the detection of gravitational waves see the contribution by Jean-Yves Vinet to this Poincaré seminar.

Let us first consider the simplest case of very weak gravitational waves, outside of their material sources. The geometry of such a space-time can be written, as in Section 6, as $g_{\mu\nu}(x) = \eta_{\mu\nu} + h_{\mu\nu}(x)$, where $h_{\mu\nu} \ll 1$. At first order in h, and outside of the source (namely in the domain where $T_{\mu\nu}(x) = 0$), the perturbation of the geometry, $h_{\mu\nu}(x)$, satisfies a homogeneous equation obtained by replacing the right-hand side of Equation (11) with zero. It can be shown that one can simplify this equation through a suitable choice of coordinate system. In a *transverse traceless* (TT) coordinate system the only non-zero components of a general gravitational wave are the spatial components h_{ij}^{TT}, $i,j = 1,2,3$ (in other words $h_{00}^{\mathrm{TT}} = 0 = h_{0i}^{\mathrm{TT}}$), and these components satisfy

$$\Box\, h_{ij}^{\mathrm{TT}} = 0\,, \;\; \partial_j\, h_{ij}^{\mathrm{TT}} = 0\,, \;\; h_{jj}^{\mathrm{TT}} = 0\,. \tag{21}$$

The first equation in (21), where the wave operator $\Box = \Delta - c^{-2}\,\partial_t^2$ appears, shows that gravitational waves (like electromagnetic waves) propagate at the speed c. If we consider for simplicity a monochromatic plane wave ($h_{ij}^{\mathrm{TT}} = \zeta_{ij}\,\exp(i\,\boldsymbol{k}\cdot\boldsymbol{x} - i\,\omega\,t) +$ complex conjugate, with $\omega = c\,|\boldsymbol{k}|$), the second equation in (21) shows that the (complex) tensor ζ_{ij} measuring the polarization of a gravitational wave only has non-zero components in the plane orthogonal to the wave's direction of propagation: $\zeta_{ij}\,k^j = 0$. Finally, the third equation in (21) shows that the polarization tensor ζ_{ij} has vanishing trace: $\zeta_{jj} = 0$. More concretely, this means that if a gravitational wave propagates in the z-direction, its polarization is described by a 2×2 matrix, $\begin{pmatrix} \zeta_{xx} & \zeta_{xy} \\ \zeta_{yx} & \zeta_{yy} \end{pmatrix}$, which is symmetric and traceless. Such a polarization matrix therefore only contains two independent (complex) components: $\zeta_+ \equiv \zeta_{xx} = -\zeta_{yy}$, and $\zeta_\times \equiv \zeta_{xy} = \zeta_{yx}$. This is the same number of independent (complex) components that an electromagnetic wave has. Indeed, in a transverse gauge, an electromagnetic wave only has spatial components A_i^T that satisfy

$$\Box A_i^T = 0 \, , \; \partial_j A_j^T = 0 \, . \tag{22}$$

As in the case above, the first equation (22) means that an electromagnetic wave propagates at the speed c, and the second equation shows that a monochromatic plane electromagnetic wave ($A_i^T = \zeta_i \exp(i \boldsymbol{k} \cdot \boldsymbol{x} - i \omega t) +$ c.c., $\omega = c \, |\boldsymbol{k}|$) is described by a (complex) polarization vector ζ_i that is orthogonal to the direction of propagation: $\zeta_j \, k^j = 0$. For a wave propagating in the z-direction such a vector only has two independent (complex) components, ζ_x and ζ_y. It is indeed the same number of components that a gravitational wave has, but we see that the two quantities measuring the polarization of a gravitational wave, $\zeta_+ = \zeta_{xx} = -\zeta_{yy}$, $\zeta_\times = \zeta_{xy} = \zeta_{yx}$ are mathematically quite different from the two quantities ζ_x, ζ_y measuring the polarization of an electromagnetic wave. However, see Section 11 below.

We have here discussed the propagation of a gravitational wave in a background space-time described by the Minkowski metric $\eta_{\mu\nu}$. One can also consider the propagation of a wave in a curved background space-time, namely by studying solutions of Einstein's equations (9) of the form $g_{\mu\nu}(x) = g_{\mu\nu}^B(x) + h_{\mu\nu}(x)$ where $h_{\mu\nu}$ is not only small, but varies on temporal and spatial scales much shorter than those of the background metric $g_{\mu\nu}^B(x)$. Such a study is necessary, for example, for understanding the propagation of gravitational waves in the cosmological Universe.

The problem of *generation* consists in searching for the connection between the tensorial amplitude $h_{ij}^{\rm TT}$ of the gravitational radiation in the radiation zone and the motion and structure of the source. If one considers the simplest case of a source that is sufficiently diffuse that it only creates waves that are everywhere weak ($g_{\mu\nu} - \eta_{\mu\nu} = h_{\mu\nu} \ll 1$), one can use the linearized approximation to Einstein's equations (9), namely Equations (11). One can solve Equations (11) by the same technique that is used to solve Maxwell's equations (12): one fixes the coordinate system by imposing $\partial^\alpha \, h_{\alpha\mu} - \frac{1}{2} \, \partial_\mu \, h_\alpha^\alpha = 0$ (analogous to the Lorentz gauge condition $\partial^\alpha \, A_\alpha = 0$), then one inverts the wave operator by using retarded potentials. Finally, one must study the asymptotic form, at infinity, of the emitted wave, and write it in the reduced form of a transverse and traceless amplitude $h_{ij}^{\rm TT}$ satisfying Equations (21) (analogous to a transverse electromagnetic wave A_i^T satisfying (22)). One then finds that, just as charge conservation implies that there is no monopole type electro-magnetic radiation, but only dipole or higher orders of polarity, the conservation of energy-momentum implies the absence of monopole *and* dipole gravitational radiation. For a slowly varying source ($v/c \ll 1$), the dominant gravitational radiation is of *quadrupole* type. It is given, in the radiation zone, by an expression of the form

$$h_{ij}^{TT}(t,r,\boldsymbol{n}) \simeq \frac{2G}{c^4\,r}\,\frac{\partial^2}{\partial t^2}\,[I_{ij}(t-r/c)]^{\mathrm{TT}}\,. \tag{23}$$

Here r denotes the distance to the center of mass of the source, $I_{ij}(t) \equiv \int d^3x\, c^{-2}\, T^{00}(t,\boldsymbol{x})\left(x^i x^j - \frac{1}{3}\,\boldsymbol{x}^2\delta^{ij}\right)$ is the quadrupole moment of the mass-energy distribution, and the upper index TT denotes an algebraic projection operation for the quadrupole tensor I_{ij} (which is a 3×3 matrix) that only retains the part orthogonal to the local direction of wave propagation $n^i \equiv x^i/r$ with vanishing trace (I_{ij}^{TT} is therefore locally a (real) 2×2 symmetric, traceless matrix of the same type as ζ_{ij} above). Formula (23) (which was in essence obtained by Einstein in 1918 [15]) is only the first approximation to an expansion in powers of v/c, where v designates an internal speed characteristic of the source. The prospect of soon being able to detect gravitational waves has motivated theorists to improve Formula (23): (i) by describing the terms of higher order in v/c, up to a very high order, and (ii) by using new approximation methods that allow one to treat sources containing regions of strong gravitational fields (such as, for example, a binary system of two black holes or two neutron stars). See below for the most recent results.

Finally, the problem of *detection*, of which the pioneer was Joseph Weber in the 1960s, is at present giving rise to very active experimental research. The principle behind any detector is that a gravitational wave of amplitude h_{ij}^{TT} induces a change in the distance L between two bodies on the order of $\delta L \sim hL$ during its passage. One way of seeing this is to consider the action of a wave h_{ij}^{TT} on two free particles, at rest before the arrival of the wave at the positions x_1^i and x_2^i respectively. As we have seen, each particle, in the presence of the wave, will follow a geodesic motion in the geometry $g_{\mu\nu} = \eta_{\mu\nu} + h_{\mu\nu}$ (with $h_{00} = h_{0i} = 0$ and $h_{ij} = h_{ij}^{\mathrm{TT}}$). By writing out the geodesic equation, Equation (7), one finds that it simply reduces (at first order in h) to $d^2x^i/ds^2 = 0$. Therefore, particles that are initially at rest ($x^i =$ const.) remain at rest in a transverse and traceless system of coordinates! This does not however mean that the gravitational wave has no observable effect. In fact, since the spatial geometry is perturbed by the passage of the wave, $g_{ij}(t,\boldsymbol{x}) = \delta_{ij} + h_{ij}^{\mathrm{TT}}(t,\boldsymbol{x})$, one finds that the physical distance between the two particles x_1^i, x_2^i (which is observable, for example, by measuring the time taken for light to make a round trip between the two particles) varies, during the passage of the wave, according to $L^2 = (\delta_{ij} + h_{ij}^{\mathrm{TT}})(x_2^i - x_1^i)(x_2^j - x_1^j)$.

The problem of detecting a gravitational wave thus leads to the problem of detecting a small relative displacement $\delta L/L \sim h$. By using Formula (23), one finds that the order of magnitude of h, for known or hoped for astrophysical sources (for example, a very close system of two neutron stars or two black holes), situated at distances such that one may hope to see several events per year ($r \gtrsim 600$ million light-years), is in fact extremely small: $h \lesssim 10^{-22}$ for signals whose characteristic frequency is around 100 Hertz. Several types of detectors have been developed since the pioneering work of J. Weber [16]. At present, the detectors that should succeed in the near future at detecting amplitudes $h \sim \delta L/L \sim 10^{-22}$ are large interferometers, of the Michelson or Fabry-Pérot type, having arms that are many kilometers in length into which a very powerful monochromatic laser beam is injected. Such terrestrial interferometric detectors presently exist in the U.S.A. (the LIGO detectors [17]), in Europe (the VIRGO [18] and GEO 600 [19] detectors) and elsewhere (such as the TAMA detector in Japan). Moreover, the international space project LISA [20], made up of an interferometer between satellites that are several million kilometers apart, should allow one to detect low frequency ($\sim$ one hundredth or one thousandth of a Hertz) gravitational waves in a dozen years or so. This collection of gravitational wave detectors promises to bring invaluable information for astronomy by opening a new "window" on the Universe that is much more transparent than the various electromagnetic (or neutrino) windows that have so greatly expanded our knowledge of the Universe in the twentieth century.

The extreme smallness of the expected gravitational signals has led a number of experimentalists to contribute, over many years, a wealth of ingenuity and know-how in order to develop technology that is sufficiently precise and trustworthy (see [17, 18, 19, 20]). To conclude, let us also mention how much concerted theoretical effort has been made, both in calculating the general relativistic predictions for gravitational waves emitted by certain sources, and in developing methods adapted to the extraction of the gravitational signal from the background noise in the detectors. For example, one of the most promising sources for terrestrial detectors is the wave train for gravitational waves emitted by a system of two black holes, and in particular the final (most intense) portion of this wave train, which is emitted during the last few orbits of the system and the final coalescence of the two black holes into a single, more massive black hole. We have seen above (see Section 9) that the finite speed of propagation of the gravitational interaction between the two bodies of a binary system gives rise to a progressive acceleration of the orbital frequency, connected to the progressive approach of the two bodies towards each other. Here we are speaking of the final stages in such a process, where the two bodies are so close that they orbit around each other in a spiral pattern that accelerates until they attain (for the final "stable" orbits) speeds that become comparable to the speed of light, all the while remaining slightly slower. In order to be able to determine, with a precision that is acceptable for the needs of detection, the dynamics of such a binary black hole system in such a situation, as well as the gravitational amplitude $h_{ij}^{\rm TT}$ that it emits, it was necessary to develop a whole ensemble of analytic techniques to a very high level of precision. For example, it was necessary to calculate the expansion (20) of the force determining the motion of the two bodies to a very high order and also to calculate the amplitude $h_{ij}^{\rm TT}$ of the gravitational radiation emitted to infinity with a precision going well beyond the quadrupole approximation (23). These calculations are comparable in complexity to high-order calculations in quantum field theory. Some of the techniques developed for quantum field theory indeed proved to be extremely useful for these calculations in the (classical) theory of general relativity (such as certain resummation methods and the mathematical use of analytic continuation in the number of space-time dimensions). For an entryway into the literature of these modern analytic methods, see [21], and for an early example of a result obtained by such methods of direct interest for the physics of detection see Figure 5 [22], which shows a component of the gravitational amplitude $h_{ij}^{\mathrm{TT}}(t)$ emitted during the final stages of evolution of a system of two black holes of equal mass. The first oscillations shown in Figure 5 are emitted during the last quasi-circular orbits (accelerated motion in a spiral of decreasing radius). The middle part of the signal corresponds to a phase where, having moved past the last stable orbit, the two black holes "fall" toward each other while spiraling rapidly. In fact, contrarily to Newton's theory, which predicts that two condensed bodies would be able to orbit around each other with an orbit of arbitrarily small radius (basically up until the point that the two bodies touch), Einstein's theory predicts a modified law for the force between the two bodies, Equation (20), whose analysis shows that it is so attractive that it no longer allows for stable circular orbits when the distance between the two bodies becomes smaller than around $6\,G(m_A + m_B)/c^2$. In the case of two black holes, this distance is sufficiently larger than the black hole "radii" ($2\,G\,m_A/c^2$ and $2\,G\,m_B/c^2$) that one is still able to analytically treat the beginning of the "spiraling plunge" of the two black holes towards each other. The final oscillations in Figure 5 are emitted by the rotating (and initially highly deformed) black hole formed from the merger of the two initial, separate black holes.

Up until quite recently the analytic predictions illustrated in Figure 5 concerning the gravitational signal $h(t)$ emitted by the spiraling plunge and merger of two black holes remained conjectural, since they could be compared to neither other theoretical predictions nor to observational data. Recently, worldwide efforts made over three decades to attack the problem of the coalescence of two black holes by *numerically* solving Einstein's equations (9) have spectacularly begun to bear fruit. Several groups have been able to numerically calculate the signal $h(t)$ emitted during the final orbits and merger of two black holes [23]. In essence, there is good agreement between the analytical and numerical predictions. In order to be able to detect the gravitational waves emitted by the coalescence of two black holes, it will most likely be necessary to properly combine the information on the structure of the signal $h(t)$ obtained by the two types of methods, which are in fact complementary.

11. General Relativity and Quantum Theory: From Supergravity to String Theory

Up until now, we have discussed the *classical* theory of general relativity, neglecting any quantum effects. What becomes of the theory in the quantum regime? This apparently innocent question in fact opens up vast new prospects that are still under construction. We will do nothing more here than to touch upon the subject,

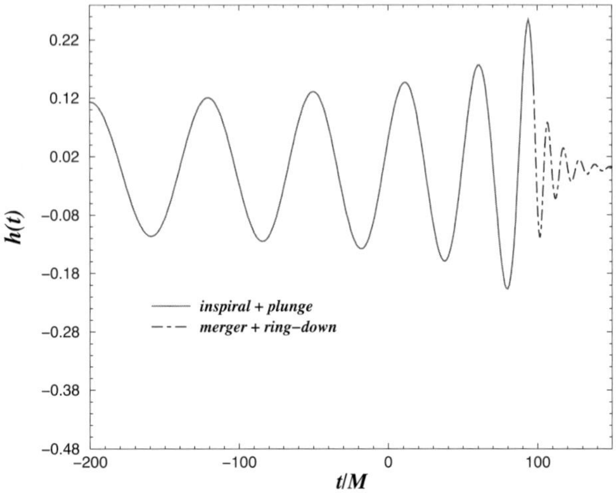

FIGURE 5. The gravitational amplitude $h(t)$ emitted during the final stages of evolution of a system of two equal-mass black holes. The beginning of the signal (the left side of the figure), which is sinusoidal, corresponds to an inspiral motion of two separate black holes (with decreasing distance); the middle corresponds to a rapid "inspiralling plunge" of the two black holes towards each other; the end (at right) corresponds to the oscillations of the final, rotating black hole formed from the merger of the two initial black holes.

by pointing out to the reader some of the paths along which contemporary physics has been led by the challenge of unifying general relativity and quantum theory. For a more complete introduction to the various possibilities "beyond" general relativity suggested within the framework of *string theory* (which is still under construction) one should consult the contribution of Ignatios Antoniadis to this Poincaré Seminar.

Let us recall that, from the very beginning of the quasi-definitive formulation of quantum theory (1925–1930), the creators of quantum mechanics (Born, Heisenberg, Jordan; Dirac; Pauli; etc.) showed how to "quantize" not only systems with several particles (such as an atom), but also *fields*, continuous dynamical systems whose classical description implies a continuous distribution of energy and momentum in space. In particular, they showed how to *quantize* (or in other words how to formulate within a framework compatible with quantum theory) the electromagnetic field A_μ, which, as we have recalled above, satisfies the Maxwell equations (12) at the classical level. They nevertheless ran into difficulty due to the following fact. In quantum theory, the physics of a system's evolution is essentially contained in the *transition amplitudes* $A(f, i)$ between an initial state labelled by i and a final state labelled by f. These amplitudes $A(f,i)$ are complex numbers. They satisfy a "transitivity" property of the type

$$A(f,i) = \sum_n A(f,n)\, A(n,i)\,, \tag{24}$$

which contains a sum over all possible intermediate states, labelled by n (with this sum becoming an integral when there is a continuum of intermediate possible states). R. Feynman used Equation (24) as a point of departure for a new formulation of quantum theory, by interpreting it as an analog of *Huygens' Principle*: if one thinks of $A(f,i)$ as the amplitude, "at the point f," of a "wave" emitted "from the point i," Equation (24) states that this amplitude can be calculated by considering the "wave" emitted from i as passing through all possible intermediate "points" n ($A(n,i)$), while re-emitting "wavelets" starting from these intermediate points ($A(f,n)$), which then superpose to form the total wave arriving at the "final point f."

Property (24) does not pose any problem in the quantum mechanics of discrete systems (particle systems). It simply shows that the amplitude $A(f,i)$ behaves like a wave, and therefore must satisfy a "wave equation" (which is indeed the case for the Schrödinger equation describing the dependence of $A(f,i)$ on the parameters determining the final configuration f). On the other hand, Property (24) poses formidable problems when one applies it to the quantization of continuous dynamical systems (fields). In fact, for such systems the "space" of intermediate possible states is infinitely larger than in the case of the mechanics of discrete systems. Roughly speaking, the intermediate possible states for a field can be described as containing $\ell = 1, 2, 3, \ldots$ quantum excitations of the field, with each quantum excitation (or pair of "virtual particles") being described essentially by a plane wave, $\zeta \exp(i\, k_\mu\, x^\mu)$, where ζ measures the polarization of these virtual particles and $k^\mu = \eta^{\mu\nu}\, k_\nu$, with $k^0 = \omega$ and $k^i = \boldsymbol{k}$, their angular frequency and wave vector, or (using the Planck-Einstein-de Broglie relations $E = \hbar\,\omega$, $\boldsymbol{p} = \hbar\,\boldsymbol{k}$) their energy-momentum $p^\mu = \hbar\, k^\mu$. The quantum theory shows (basically because of the uncertainty principle) that the four-frequencies (and four-momenta) $p^\mu = \hbar\, k^\mu$ of the intermediate states cannot be constrained to satisfy the classical equation $\eta_{\mu\nu}\, p^\mu\, p^\nu = -m^2$ (or in other words $E^2 = \boldsymbol{p}^2 + m^2$; we use $c = 1$ in this section). As a consequence, the sum over intermediate states for a quantum field theory has the following properties (among others): (i) when $\ell = 1$ (an intermediate state containing only one pair of virtual particles, called a *one-loop contribution*), there is an integral over a four-momentum p^μ, $\int d^4p = \int dE \int d\boldsymbol{p}$; (ii) when $\ell = 2$ (two pairs of virtual particles; a *two-loop contribution*), there is an integral over two four-momenta p_1^μ, p_2^μ, $\int d^4p_1\, d^4p_2$; etc. The delicate point comes from the fact that the energy-momentum of an intermediate state can take arbitrarily high values. This possibility is directly connected (through a Fourier transform) to the fact that a field possesses an infinite number of degrees of freedom, corresponding to configurations that vary over arbitrarily small time and length scales.

The problems posed by the necessity of integrating over the infinite domain of four-momenta of intermediate virtual particles (or in other words of accounting for the fact that field configurations can vary over arbitrarily small scales) appeared in the 1930s when the quantum theory of the electromagnetic field A_μ (called quantum electrodynamics, or QED) was studied in detail. These problems imposed themselves in the following form: when one calculates the transition amplitude for given initial and final states (for example the collision of two light quanta, with two photons entering and two photons leaving) by using (24), one finds a result given in the form of a *divergent integral*, because of the integral (in the one-loop approximation, $\ell = 1$) over the arbitrarily large energy-momentum describing virtual electron-positron pairs appearing as possible intermediate states. Little by little, theoretical physicists understood that the types of divergent integrals appearing in QED were relatively benign and, after the second world war, they developed a method (*renormalization theory*) that allowed one to unambiguously isolate the infinite part of these integrals, and to subtract them by expressing the amplitudes $A(f, i)$ solely as a function of observable quantities [24] (work by J. Schwinger, R. Feynman, F. Dyson etc.).

The preceding work led to the development of consistent quantum theories not only for the electromagnetic field A_μ (QED), but also for generalizations of electromagnetism (Yang-Mills theory or non-abelian gauge theory) that turned out to provide excellent descriptions of the new interactions between elementary particles discovered in the twentieth century (the electroweak theory, partially unifying electromagnetism and weak nuclear interactions, and quantum chromodynamics, describing the strong nuclear interactions). All of these theories give rise to only relatively benign divergences that can be "renormalized" and thus allowed one to compute amplitudes $A(f, i)$ corresponding to observable physical processes [24] (notably, work by G. 't Hooft and M. Veltman).

What happens when we use (24) to construct a "perturbative" quantum theory of general relativity (namely one obtained by expanding in the number ℓ of virtual particle pairs appearing in the intermediate states)? The answer is that the integrals over the four-momenta of intermediate virtual particles are not at all of the benign type that allowed them to be renormalized in the simpler case of electromagnetism. The source of this difference is not accidental, but is rather connected with the basic physics of relativistic gravitation. Indeed, as we have mentioned, the virtual particles have arbitrarily large energies E. Because of the basic relations that led Einstein to develop general relativity, namely $E = m_i$ and $m_i = m_g$, one deduces that these virtual particles correspond to arbitrarily large gravitational masses m_g. They will therefore end up creating intense gravitational effects that become more and more intense as the number ℓ of virtual particle pairs grows. These gravitational interactions that grow without limit with energy and momentum correspond (by Fourier transform) to field configurations concentrated in arbitrarily small space and time scales. One way of seeing why the quantum gravitational field creates much more violent problems than the quantum electromagnetic field is, quite simply, to go back to dimensional analysis. Simple considerations in fact show that the relative (non-dimensional) one-loop amplitude A_1 must be proportional to the product $\hbar G$ and must contain an integral $\int d^4k$. However, in 1900 Planck had noticed that (in units where $c = 1$) the dimensions of $\hbar$ and G were such that the product $\hbar G$ had the dimensions of length (or time) squared:

$$\ell_P \equiv \sqrt{\frac{\hbar G}{c^3}} \simeq 1.6 \times 10^{-33}\,\mathrm{cm}, \ t_P \equiv \sqrt{\frac{\hbar G}{c^5}} \simeq 5.4 \times 10^{-44}\,\mathrm{s}\,. \tag{25}$$

One thus deduces that the integral $\int d^4k\, f(k)$ must have the dimensions of a squared frequency, and therefore that A_1 must (when $k \to \infty$) be of the type, $A_1 \sim \hbar G \int d^4k/k^2$. Such an integral diverges quadratically with the upper limit Λ of the integral (the cutoff frequency, such that $|k| \leq \Lambda$), so that $A_1 \sim \hbar G \Lambda^2 \sim t_P^2 \Lambda^2$. The extension of this dimensional analysis to the intermediate states with several loops ($\ell > 1$) causes even more severe polynomial divergences to appear, of a type such that the power of Λ that appears grows without limit with ℓ.

In summary, the essential physical characteristics of gravitation ($E = m_i = m_g$ and the dimension of Newton's constant G) imply the impossibility of generalizing to the gravitational case the methods that allowed a satisfactory quantum treatment of the other interactions (electromagnetic, weak, and strong). Several paths have been explored to get out of this impasse. Some researchers tried to quantize general relativity non-perturbatively, without using an expansion in intermediate states (24) (work by A. Ashtekar, L. Smolin, and others). Others have tried to generalize general relativity by adding a fermionic field to Einstein's (bosonic) gravitational field $g_{\mu\nu}(x)$, the gravitino field $\psi_\mu(x)$. It is indeed remarkable that it is possible to define a theory, known as *supergravity*, that generalizes the geometric invariance of general relativity in a profound way. After the 1974 discovery (by J. Wess and B. Zumino) of a possible new global symmetry for interacting bosonic and fermionic fields, *supersymmetry* (which is a sort of global rotation transforming bosons to fermions and vice versa), D.Z. Freedman, P. van Nieuwenhuizen, and S. Ferrara; and S. Deser and B. Zumino; showed that one could generalize global supersymmetry to a *local supersymmetry*, meaning that it varies from point to point in space-time. Local supersymmetry is a sort of fermionic generalization (with anti-commuting parameters) of the geometric invariance at the base of general relativity (the invariance under any change in coordinates). The generalization of Einstein's theory of gravitation that admits such a local supersymmetry is called *supergravity theory*. As we have mentioned, in four dimensions this theory contains, in addition to the (commuting) bosonic field $g_{\mu\nu}(x)$, an (anti-commuting) fermionic field $\psi_\mu(x)$ that is both a space-time vector (with index μ) and a spinor. (It is a massless field of spin 3/2, intermediate between a massless spin 1 field like A_μ and a massless spin 2 field like $h_{\mu\nu} = g_{\mu\nu} - \eta_{\mu\nu}$.) Supergravity was extended to richer fermionic structures (with many gravitinos), and was formulated in space-times having more than four dimensions. It is nevertheless remarkable that there is a maximal dimension, equal to $D = 11$, admitting a theory of supergravity (the maximal supergravity constructed by E. Cremmer, B. Julia, and J. Scherk). The initial hope underlying the construction of these supergravity theories was that they would perhaps allow one to give meaning to the perturbative calculation (24) of quantum amplitudes. Indeed, one finds for example that at one loop, $\ell = 1$, the contributions coming from intermediate fermionic states have a sign opposite to the bosonic contributions and (because of the supersymmetry, bosons $\leftrightarrow$ fermions) exactly cancel them. Unfortunately, although such cancellations exist for the lowest orders of approximation, it appeared that this was probably not going to be the case at all orders[5]. The fact that the gravitational interaction constant G has "a bad dimension" remains true and creates non-renormalizable divergences starting at a certain number of loops ℓ.

Meanwhile, a third way of defining a consistent quantum theory of gravity was developed, under the name of *string theory*. Initially formulated as models for the strong interactions (in particular by G. Veneziano, M. Virasoro, P. Ramond, A. Neveu, and J.H. Schwarz), the string theories were founded upon the quantization of the relativistic dynamics of an extended object of one spatial dimension: a "string." This string could be closed in on itself, like a small rubber band (a closed string), or it could have two ends (an open string). Note that the point of departure of string theory only includes the Poincaré-Minkowski space-time, in other words the metric $\eta_{\mu\nu}$ of Equation (2), and quantum theory (with the constant $\hbar = h/2\pi$). In particular, the only symmetry manifest in the classical dynamics of a string is the Poincaré group (3). It is, however, remarkable that (as shown by T. Yoneya, J. Scherk and J.H. Schwarz, in 1974) one of the quantum excitations of a closed string reproduces, in a certain limit, all of the non-linear structure of general relativity (see below). Among the other remarkable properties of string theory [25], let us point out that it is the first physical theory to determine the space-time dimension D. In fact, this theory is only consistent if $D = 10$, for the versions allowing fermionic excitations (the purely bosonic string theory selects $D = 26$). The fact that $10 > 4$ does not mean that this theory has no relevance to the real world. Indeed, it has been known since the 1930s (from work of T. Kaluza and O. Klein) that a space-time of dimension $D > 4$ is compatible with experiment if the supplementary (spatial) dimensions close in on themselves (meaning they are *compactified*) on very small distance scales. The low-energy physics of such a theory seems to take place in a four-dimensional space-time, but it contains new (a priori massless) fields connected to the geometry of the additional compactified dimensions. Moreover, recent work (due in particular to I. Antoniadis, N. Arkani-Hamed, S. Dimopoulos, and G. Dvali) has suggested the possibility that the additional dimensions are compactified on scales that are small with respect to everyday life, but very large with respect to the Planck length. This possibility opens up an entire phenomenological field dealing with the eventual observation

[5]Recent work by Z. Bern et al. and M. Green et al., has, however, suggested that such cancellations take place at all orders for the case of maximal supergravity, dimensionally reduced to $D = 4$ dimensions.

of signals coming from string theory (see the contribution of I. Antoniadis to this Poincaré seminar).

However, string theory's most remarkable property is that it seems to avoid, in a radical way, the problems of divergent (non-renormalizable) integrals that have weighed down every direct attempt at perturbatively quantizing gravity. In order to explain how string theory arrives at such a result, we must discuss some elements of its formalism.

Recall that the classical dynamics of any system is obtained by minimizing a functional of the time evolution of the system's configuration, called the *action* (the principle of least action). For example, the action for a particle of mass m, moving in a Riemannian space-time (6), is proportional to the length of the line that it traces in space-time: $S = -m \int ds$. This action is minimized when the particle follows a geodesic, in other words when its equation of motion is given by (7). According to Y. Nambu and T. Goto, the action for a string is $S = -T \iint dA$, where the parameter T (analogous to m for the particle) is called the string *tension*, and where $\iint dA$ is the area of the *two-dimensional* surface traced out by the evolution of the string in the (D-dimensional) space-time in which it lives. In quantum theory, the action functional serves (as shown by R. Feynman) to define the transition amplitude (24). Basically, when one considers two intermediate configurations m and n (in the sense of the right-hand side of (24)) that are close to each other, the amplitude $A(n,m)$ is proportional to $\exp(i\, S(n,m)/\hbar)$, where $S(n,m)$ is the minimal classical action such that the system considered evolves from the configuration labelled by n to that labelled by m. Generalizing the decomposition in (24) by introducing an infinite number of intermediate configurations that lie close to each other, one ends up (in a generalization of Huygens' principle) expressing the amplitude $A(f,i)$ as a multiple sum over all of the "paths" (in the configuration space of the system studied) connecting the initial state i to the final state f. Each path contributes a term $e^{i\phi}$ where the phase $\phi = S/\hbar$ is proportional to the action S corresponding to this "path", or in other words to this possible evolution of the system. In string theory, $\phi = -(T/\hbar) \iint dA$. Since the phase is a non-dimensional quantity, and $\iint dA$ has the dimension of an area, we see that the quantum theory of strings brings in the quantity $\hbar/T$, having the dimensions of a length squared, at a fundamental level. More precisely, the fundamental length of string theory, ℓ_s, is defined by

$$\ell_s^2 \equiv \alpha' \equiv \frac{\hbar}{2\pi T}\,. \tag{26}$$

This fundamental length plays a central role in string theory. Roughly speaking, it defines the characteristic "size" of the quantum states of a string. If ℓ_s is much smaller than the observational resolution with which one studies the string, the string will look like a point-like particle, and its interactions will be described by a quantum theory of relativistic particles, which is equivalent to a theory of relativistic fields. It is precisely in this sense that general relativity emerges as a limit of string theory. Since this is an important conceptual point for our story, let us give some details about the emergence of general relativity from string theory.

The action functional that is used in practice to quantize a string is not really $-T\iint dA$, but rather (as emphasized by A. Polyakov)

$$\frac{S}{\hbar} = -\frac{1}{4\,\pi\,\ell_s^2}\iint d^2\sigma\,\sqrt{-\gamma}\,\gamma^{ab}\,\partial_a\,X^\mu\,\partial_b\,X^\nu\,\eta_{\mu\nu}+\cdots\,, \tag{27}$$

where σ^a, $a=0,1$ are two coordinates that allow an event to be located on the space-time surface (or 'world-sheet') traced out by the string within the ambient space-time; γ_{ab} is an auxiliary metric ($d\,\Sigma^2 = \gamma_{ab}(\sigma)\,d\sigma^a\,d\sigma^b$) defined on this surface (with γ^{ab} being its inverse, and γ its determinant); and $X^\mu(\sigma^a)$ defines the embedding of the string in the ambient (flat) space-time. The dots indicate additional terms, and in particular terms of fermionic type that were introduced by P. Ramond, by A. Neveu and J.H. Schwarz, and by others. If one separates the two coordinates $\sigma^a = (\sigma^0,\sigma^1)$ into a temporal coordinate, $\tau\equiv\sigma^0$, and a spatial coordinate, $\sigma\equiv\sigma^1$, the configuration "at time τ" of the string is described by the functions $X^\mu(\tau,\sigma)$, where one can interpret σ as a curvilinear abscissa describing the spatial extent of the string. If we consider a closed string, one that is topologically equivalent to a circle, the function $X^\mu(\tau,\sigma)$ must be periodic in σ. One can show that (modulo the imposition of certain constraints) one can choose the coordinates τ and σ on the string such that $d\,\Sigma^2 = -d\tau^2+d\sigma^2$. Then, the dynamical equations for the string (obtained by minimizing the action (27)) reduce to the standard equation for waves on a string: $-\partial^2X^\mu/\partial\tau^2+\partial^2X^\mu/\partial\sigma^2=0$. The general solution to this equation describes a superposition of waves travelling along the string in both possible directions: $X^\mu = X_L^\mu(\tau+\sigma)+X_R^\mu(\tau-\sigma)$. If we consider a closed string (one that is topologically equivalent to a circle), these two types of wave are independent of each other. For an open string (with certain reflection conditions at the endpoints of the string) these two types of waves are connected to each other. Moreover, since the string has a finite length in both cases, one can decompose the left- or right-moving waves $X_L^\mu(\tau+\sigma)$ or $X_R^\mu(\tau-\sigma)$ as a Fourier series. For example, for a closed string one may write

$$X^\mu(\tau,\sigma) = X_0^\mu(\tau)+\frac{i}{\sqrt{2}}\,\ell_s\sum_{n=1}^{\infty}\left(\frac{a_n^\mu}{\sqrt{n}}\,e^{-2in(\tau-\sigma)}+\frac{\tilde a_n^\mu}{\sqrt{n}}\,e^{-2in(\tau+\sigma)}\right)+\text{h.c.} \tag{28}$$

Here $X_0^\mu(\tau) = x^\mu + 2\,\ell_s^2\,p^\mu\tau$ describes the motion of the string's center of mass, and the remainder describes the decomposition of the motion around the center of mass into a discrete set of oscillatory modes. Like any vibrating string, a relativistic string can vibrate in its fundamental mode ($n=1$) or in a "harmonic" of the fundamental mode (for an integer $n>1$). In the classical case the complex coefficients a_n^μ, $\tilde a_n^\mu$ represent the (complex) amplitudes of vibration for the modes of oscillation at frequency n times the fundamental frequency. (with a_n^μ corresponding to a wave travelling to the right, while $\tilde a_n^\mu$ corresponds to a wave travelling to the left.) When one quantizes the string dynamics the position of the string $X^\mu(\tau,\sigma)$ becomes an operator (acting in the space of quantum states of the system), and because of this the quantities x^μ, p^μ, a_n^μ and $\tilde a_n^\mu$ in (28) become operators. The notation h.c. signifies that one must add the hermitian conjugates of the oscillation terms, which will contain the operators $(a_n^\mu)^\dagger$ and $(\tilde{a}_n^\mu)^\dagger$. (The notation $\dagger$ indicates hermitian conjugation, in other words the operator analog of complex conjugation.) One then finds that the operators x^μ and p^μ describing the motion of the center of mass satisfy the usual commutation relations of a relativistic particle, $[x^\mu, p^\mu] = i\,\hbar\,\eta^{\mu\nu}$, and that the operators a_n^μ and $\tilde{a}_n^\mu$ become annihilation operators, like those that appear in the quantum theory of any vibrating system: $[a_n^\mu, (a_m^\nu)^\dagger] = \hbar\,\eta^{\mu\nu}\,\delta_{nm}$, $[\tilde{a}_n^\mu, (\tilde{a}_m^\nu)^\dagger] = \hbar\,\eta^{\mu\nu}\,\delta_{mn}$. In the case of an open string, one only has *one* set of oscillators, let us say a_n^μ. The discussion up until now has neglected to mention that the oscillation amplitudes $a_n^\mu, \tilde{a}_n^\mu$ must satisfy an infinite number of constraints (connected with the equation obtained by minimizing (27) with respect to the auxiliary metric γ_{ab}). One can satisfy these by expressing *two* of the space-time components of the oscillators $a_n^\mu, \tilde{a}_n^\mu$ (for each n) as a function of the other. Because of this, the physical states of the string are described by oscillators $a_n^i, \tilde{a}_n^i$ where the index i only takes $D-2$ values in a space-time of dimension D. Forgetting this subtlety for the moment (which is nevertheless crucial physically), let us conclude this discussion by summarizing the *spectrum* of a quantum string, or in other words the ensemble of quantum states of motion for a string.

For an open string, the ensemble of quantum states describes the states of motion (the momenta p^μ) of an infinite collection of relativistic particles, having squared masses $M^2 = -\eta_{\mu\nu}\,p^\mu\,p^\nu$ equal to $(N-1)\,m_s^2$, where N is a non-negative integer and $m_s \equiv \hbar/\ell_s$ is the fundamental mass of string theory associated to the fundamental length ℓ_s. For a closed string, one finds another "infinite tower" of more and more massive particles, this time with $M^2 = 4(N-1)\,m_s^2$. In both cases the integer N is given, as a function of the string's oscillation amplitudes (travelling to the right), by

$$N = \sum_{n=1}^{\infty} n\,\eta_{\mu\nu}(a_n^\mu)^\dagger\,a_n^\nu\,. \tag{29}$$

In the case of a closed string one must also satisfy the constraint $N = \tilde{N}$ where $\tilde{N}$ is the operator obtained by replacing a_n^μ by $\tilde{a}_n^\mu$ in (29).

The preceding result essentially states that the (quantized) internal energy of an oscillating string defines the squared mass of the associated particle. The presence of the additional term -1 in the formulae given above for M^2 means that the quantum state of minimum internal energy for a string, that is, the "vacuum" state $|0\rangle$ where all oscillators are in their ground state, $a_n^\mu \mid 0\rangle = 0$, corresponds to a negative squared mass ($M^2 = -m_s^2$ for the open string and $M^2 = -4\,m_s^2$ for the closed string). This unusual quantum state (a *tachyon*) corresponds to an instability of the theory of bosonic strings. It is absent from the more sophisticated versions of string theory ("superstrings") due to F. Gliozzi, J. Scherk, and D. Olive, to M. Green and J.H. Schwarz, and to D. Gross and collaborators. Let us concentrate on the other states (which are the only ones that have corresponding states in superstring theory). One then finds that the first possible physical quantum states (such that $N = 1$) describe some massless particles. In relativistic quantum theory it is known that any particle is the quantized excitation of a corresponding field. Therefore the massless particles that appear in string theory must correspond to long-range fields. To know which fields appear in this way one must more closely examine which possible combinations of oscillator excitations $a_1^\mu, a_2^\mu, a_3^\mu, \ldots$, appearing in Formula (29), can lead to $N = 1$. Because of the factor n in (29) multiplying the harmonic contribution of order n to the mass squared, only the oscillators of the fundamental mode $n = 1$ can give $N = 1$. One then deduces that the internal quantum states of massless particles appearing in the theory of *open strings* are described by a string oscillation state of the form

$$\zeta_\mu (a_1^\mu)^\dagger \mid 0\rangle \,. \tag{30}$$

On the other hand, because of the constraint $N = \tilde{N} = 1$, the internal quantum states of the massless particles appearing in the theory of *closed strings* are described by a state of excitation containing both a left-moving oscillation and a right-moving oscillation:

$$\zeta_{\mu\nu} (a_1^\mu)^\dagger \, (\tilde{a}_1^\nu)^\dagger \mid 0\rangle \,. \tag{31}$$

In Equations (30) and (31) the state $|0\rangle$ denotes the ground state of all oscillators $(a_n^\mu \mid 0\rangle = \tilde{a}_n^\mu \mid 0\rangle = 0)$.

The state (30) therefore describes a massless particle (with momentum satisfying $\eta_{\mu\nu}\, p^\mu\, p^\nu = 0$), possessing an "internal structure" described by a vector polarization ζ_μ. Here we recognize exactly the definition of a photon, the quantum state associated with a wave $A_\mu(x) = \zeta_\mu \exp(i\, k_\lambda\, x^\lambda)$, where $p^\mu = \hbar\, k^\mu$. The theory of open strings therefore contains Maxwell's theory. (One can also show that, because of the constraints briefly mentioned above, the polarization ζ_μ must be transverse, $k^\mu\, \zeta_\mu = 0$, and that it is only defined up to a gauge transformation: $\zeta'_\mu = \zeta_\mu + a\, k_\mu$.) As for the state (31), this describes a massless particle ($\eta_{\mu\nu}\, p^\mu\, p^\nu = 0$), possessing an "internal structure" described by a tensor polarization $\zeta_{\mu\nu}$. The plane wave associated with such a particle is therefore of the form $\bar{h}_{\mu\nu}(x) = \zeta_{\mu\nu} \exp(i\, k_\lambda\, x^\lambda)$, where $p^\mu = \hbar\, k^\mu$. As in the case of the open string, one can show that $\zeta_{\mu\nu}$ must be transverse, $\zeta_{\mu\nu}\, k^\nu = 0$ and that it is only defined up to a gauge transformation, $\zeta'_{\mu\nu} = \zeta_{\mu\nu} + k_\mu\, a_\nu + k_\nu\, b_\mu$. We here see the same type of structure appear that we had in general relativity for plane waves. However, here we have a structure that is richer than that of general relativity. Indeed, since the state (31) is obtained by combining two independent states of oscillation, $(a_1^\mu)^\dagger$ and $(\tilde{a}_1^\mu)^\dagger$, the polarization tensor $\zeta_{\mu\nu}$ is not constrained to be symmetric. Moreover it is not constrained to have vanishing trace. Therefore, if we decompose $\zeta_{\mu\nu}$ into its possible irreducible parts (a symmetric traceless part, a symmetric part with trace, and an antisymmetric part) we find that the field $\bar{h}_{\mu\nu}(x)$ associated with the massless states of a closed string decomposes into: (i) a field $h_{\mu\nu}(x)$ (the *graviton*) representing a weak gravitational wave in general relativity, (ii) a scalar field $\Phi(x)$ (called the *dilaton*), and (iii) an antisymmetric tensor field $B_{\mu\nu}(x) = -B_{\nu\mu}(x)$ subject to the gauge invariance $B'_{\mu\nu}(x) = B_{\mu\nu}(x) + \partial_\mu\, a_\nu(x) - \partial_\nu\, a_\mu(x)$. Moreover, when one studies the non-linear interactions between these various fields, as described by the transition amplitudes $A(f, i)$ in string theory, one can show that the field $h_{\mu\nu}(x)$ truly represents a deformation of the flat geometry of the background space-time in which the theory was initially formulated. Let us emphasize this remarkable result. We started from a theory that studied the quantum dynamics of a string in a rigid background space-time. This theory predicts that *certain quantum excitations of a string* (that propagate at the speed of light) *in fact represent waves of deformation of the space-time geometry.* In intuitive terms, the "elasticity" of space-time postulated by the theory of general relativity appears here as being due to certain internal vibrations of an elastic object extended in one spatial dimension.

Another suggestive consequence of string theory is the link suggested by the comparison between (30) and (31). Roughly, Equation (31) states that the internal state of a closed string corresponding to a graviton is constructed by taking the (tensor) product of the states corresponding to photons in the theory of open strings. This unexpected link between Einstein's gravitation ($g_{\mu\nu}$) and Maxwell's theory (A_μ) translates, when we look at interactions in string theory, into remarkable identities (due to H. Kawai, D.C. Lewellen, and S.-H.H. Tye) between the transition amplitudes of open strings and those of closed strings. This affinity between electromagnetism, or rather Yang-Mills theory, and gravitation has recently given rise to fascinating conjectures (due to A. Polyakov and J. Maldacena) connecting quantum Yang-Mills theory in flat space-time to quasi-classical limits of string theory and gravitation in curved space-time. Einstein would certainly have been interested to see how classical general relativity is used here to clarify the limit of a *quantum* Yang-Mills theory.

Having explained the starting point of string theory, we can outline the intuitive reason for which this theory avoids the problems with divergent integrals that appeared when one tried to directly quantize gravitation. We have seen that string theory contains an infinite tower of particles whose masses grow with the degree of excitation of the string's internal oscillators. The gravitational field appears in the limit that one considers the low energy interactions ($E \ll m_s$) between the massless states of the theory. In this limit the graviton (meaning the particle associated with the gravitational field) is treated as a "point-like" particle. When we consider more complicated processes (at one loop, $\ell = 1$, see above), virtual elementary gravitons could appear with arbitrarily high energy. It is these virtual high-energy gravitons that are responsible for the divergences. However, in string theory, when we consider any intermediate process whatsoever where high energies appear, it must be remembered that this high intermediate energy can also be used to excite the internal state of the virtual gravitons, and thus reveal that they are "made" from an extended string. An analysis of this fact shows that string theory introduces an effective truncation of the type $E \lesssim m_s$ on the energies of exchanged virtual particles. In other words, the fact that there are no truly "point-like" particles in string theory, but only string excitations having a characteristic length $\sim \ell_s$, eliminates the problem of infinities connected to arbitrarily small length and time scales. Because of this, in string theory one can calculate the transition amplitudes corresponding to a collision between two gravitons, and one finds that the result is given by a finite integral [25].

Up until now we have only considered the starting point of string theory. This is a complex theory that is still in a stage of rapid development. Let us briefly sketch some other aspects of this theory that are relevant for this exposé centered around relativistic gravitation. Let us first state that the more sophisticated versions of string theory (*superstrings*) require the inclusion of fermionic oscillators b_n^μ, $\tilde{b}_n^\mu$, in addition to the bosonic oscillators a_n^μ, $\tilde{a}_n^\mu$ introduced above. One then finds that there are no particles of negative mass-squared, and that the space-time dimension D must be equal to 10. One also finds that the massless states contain more states than those indicated above. In fact, one finds that the fields corresponding to these states describe the various possible theories of supergravity in $D = 10$. Recently (in work by J. Polchinski) it has also been understood that string theory contains not only the states of excitation of strings (in other words of objects extended in one spatial direction), but also the states of excitation of objects extended in p spatial directions, where the integer p can take other values than 1. For example, $p = 2$ corresponds to a *membrane*. It even seems (according to C. Hull and P. Townsend) that one should recognize that there is a sort of "democracy" between several different values for p. An object extended in p spatial directions is called a *p-brane*. In general, the masses of the quantum states of these p-branes are very large, being parametrically higher than the characteristic mass m_s. However, one may also consider a limit where the mass of certain p-branes tends towards zero. In this limit, the fields associated with these p-branes become long-range fields. A surprising result (by E. Witten) is that, in this limit, the infinite tower of states of certain p-branes (in particular for $p = 0$) corresponds exactly to the infinite tower of states that appear when one considers the maximal supergravity in $D = 11$ dimensions, with the eleventh (spatial) dimension compactified on a circle (that is to say with a periodicity condition on x^{11}). In other words, in a certain limit, a theory of superstrings in $D = 10$ transforms into a theory that lives in $D = 11$ dimensions! Because of this, many experts in string theory believe that the true definition of string theory (which is still to be found) must start from a theory (to be defined) in 11 dimensions (known as "M-theory").

We have seen in Section 8 that one point of contact between relativistic gravitation and quantum theory is the phenomenon of thermal emission from black holes discovered by S.W. Hawking. String theory has shed new light upon this phenomenon, as well as on the concept of black hole "entropy." The essential question that the calculation of S.W. Hawking left in the shadows is: what is the physical meaning of the quantity S defined by Equation (19)? In the thermodynamic theory of ordinary bodies, the entropy of a system is interpreted, since Boltzmann's work, as the (natural) logarithm of the number of microscopic states N having the same macroscopic characteristics (energy, volume, etc.) as the state of the system under consideration: $S = \log N$. Bekenstein had attempted to estimate the number of microscopic internal states of a macroscopically defined black hole, and had argued for a result such that $\log N$ was on the order of magnitude of $A/\hbar G$, but his arguments remained indirect and did not allow a clear meaning to be attributed to this counting of microscopic states. Work by A. Sen and by A. Strominger and C. Vafa, as well as by C.G. Callan and J.M. Maldacena has, for the first time, given examples of black holes whose microscopic description in string theory is sufficiently precise to allow for the calculation (in certain limits) of the number of internal quantum states, N. It is therefore quite satisfying to find a final result for N whose logarithm is *precisely* equal to the expression (19). However, there do remain dark areas in the understanding of the quantum structure of black holes. In particular, the string theory calculations allowing one to give a precise statistical meaning to the entropy (19) deal with very special black holes (known as *extremal* black holes, which have the maximal electric charge that a black hole with a regular horizon can support). These black holes have a Hawking temperature equal to zero, and therefore do not emit thermal radiation. They correspond to *stable* states in the quantum theory. One would nevertheless also like to understand the detailed internal quantum structure of *unstable* black holes, such as the Schwarzschild black hole (17), which has a non-zero temperature, and which therefore loses its mass little by little in the form of thermal radiation. What is the final state to which this gradual process of black hole "evaporation" leads? Is it the case that an initial pure quantum state radiates all of its initial mass to transform itself entirely into incoherent thermal radiation? Or does a Schwarzschild black hole transform itself, after having obtained a minimum size, into something else? The answers to these questions remain open to a large extent, although it has been argued that a Schwarzschild black hole transforms itself into a highly massive quantum string state when its radius becomes on the order of ℓ_s [26].

We have seen previously that string theory contains general relativity in a certain limit. At the same time, string theory is, strictly speaking, infinitely richer than Einstein's gravitation, for the graviton is nothing more than a particular quantum excitation of a string, among an infinite number of others. What deviations from Einstein's gravity are predicted by string theory? This question remains open today because of our lack of comprehension about the connection between string theory and the reality observed in our everyday environment (4-dimensional space-time; electromagnetic, weak, and strong interactions; the spectrum of observed particles; ...). We shall content ourselves here with outlining a few possibilities. (See the contribution by I. Antoniadis for a discussion of other possibilities.) First, let us state that if one considers collisions between gravitons with energy-momentum k smaller than, but not negligible with respect to, the characteristic string mass m_s, the calculations of transition amplitudes in string theory show that the usual Einstein equations (in the absence of matter) $R_{\mu\nu} = 0$ must be modified, by including corrections of order $(k/m_s)^2$. One finds that these modified Einstein equations have the form (for bosonic string theory)

$$R_{\mu\nu} + \frac{1}{4}\,\ell_s^2\,R_{\mu\alpha\beta\gamma}\,R_{\nu}^{\cdot\alpha\beta\gamma} + \cdots = 0\,, \tag{32}$$

where

$$R^{\mu}_{\cdot\nu\alpha\beta} \equiv \partial_\alpha \Gamma^{\mu}_{\nu\beta} + \Gamma^{\mu}_{\sigma\alpha}\Gamma^{\sigma}_{\nu\beta} - \partial_\beta \Gamma^{\mu}_{\nu\alpha} - \Gamma^{\mu}_{\sigma\beta}\Gamma^{\sigma}_{\nu\alpha}, \tag{33}$$

denotes the "curvature tensor" of the metric $g_{\mu\nu}$. (The quantity $R_{\mu\nu}$ defined in Section 5 that appears in Einstein's equations in an essential way is a "trace" of this tensor: $R_{\mu\nu} = R^{\sigma}_{\cdot\mu\sigma\nu}$.) As indicated by the dots in (32), the terms written are no more than the two first terms of an infinite series in growing powers of $\ell_s^2 \equiv \alpha'$. Equation (32) shows how the fact that the string is not a point, but is rather extended over a characteristic length $\sim \ell_s$, modifies the Einsteinian description of gravity. The corrections to Einstein's equation shown in (32) are nevertheless completely negligible in most applications of general relativity. In fact, it is expected that ℓ_s is on the order of the Planck scale ℓ_p, Equation (25). More precisely, one expects that ℓ_s is on the order of magnitude of 10^{-32} cm. (Nevertheless, this question remains open, and it has been recently suggested that ℓ_s is much larger, and perhaps on the order of 10^{-17} cm.)

If one assumes that ℓ_s is on the order of magnitude of 10^{-32} cm (and that the extra dimensions are compactified on distance scales on the order of ℓ_s), the only area of general relativistic applications where the modifications shown in (32) should play an important role is in primordial cosmology. Indeed, close to the initial singularity of the Big Bang (if it exists), the "curvature" $R_{\mu\nu\alpha\beta}$ becomes extremely large. When it reaches values comparable to ℓ_s^{-2} the infinite series of corrections in (32) begins to play a role comparable to the first term, discovered by Einstein. Such a situation is also found in the interior of a black hole, when one gets very close to the singularity (see Figure 3). Unfortunately, in such situations, one must take the infinite series of terms in (32) into account, or in other words replace Einstein's description of gravitation in terms of a *field* (which corresponds to a *point-like* (quantum) particle) by its exact stringy description. This is a difficult problem that no one really knows how to attack today.

However, a priori string theory predicts more drastic low energy ($k \ll m_s$) modifications to general relativity than the corrections shown in (32). In fact, we have seen in Equation (31) above that Einsteinian gravity does not appear alone in string theory. It is always necessarily accompanied by other long-range fields, in particular a scalar field $\Phi(x)$, the *dilaton*, and an antisymmetric tensor $B_{\mu\nu}(x)$. What role do these "partners" of the graviton play in observable reality? This question does not yet have a clear answer. Moreover, if one recalls that (super)string theory must live in a space-time of dimension $D = 10$, and that it includes the $D = 10$ (and eventually the $D = 11$) theory of supergravity, there are many other supplementary fields that add themselves to the ten components of the usual metric tensor $g_{\mu\nu}$ (in $D = 4$). It is conceivable that all of these supplementary fields (which are massless to first approximation in string theory) acquire masses in our local universe that are large enough that they no longer propagate observable effects over macroscopic scales. It remains possible, however, that one or several of these fields remain (essentially) massless, and therefore can propagate physical effects over distances that are large enough to be observable. It is therefore of interest to understand what physical effects are implied, for example, by the dilaton $\Phi(x)$ or by $B_{\mu\nu}(x)$. Concerning the latter, it is interesting to note that (as emphasized by A. Connes, M. Douglas, and A. Schwartz), in a certain limit, the presence of a background $B_{\mu\nu}(x)$ has the effect of deforming the space-time geometry in a "non-commutative" way. This means that, in a certain sense, the space-time coordinates x^μ cease to be simple real (commuting) numbers in order to become non-commuting quantities: $x^\mu x^\nu - x^\nu x^\mu = \varepsilon^{\mu\nu}$ where $\varepsilon^{\mu\nu} = -\varepsilon^{\nu\mu}$ is connected to a (uniform) background $B_{\mu\nu}$. To conclude, let us consider the other obligatory partner of the graviton $g_{\mu\nu}(x)$, the dilaton $\Phi(x)$. This field plays a central role in string theory. In fact, the average value of the dilaton (in the vacuum) determines the string theory coupling constant, $g_s = e^\Phi$. The value of g_s in turn determines (along with other fields) the physical coupling constants. For example, the gravitational coupling constant is given by a formula of the type $\hbar G = \ell_s^2(g_s^2 + \cdots)$ where the dots denote correction terms (which can become quite important if g_s is not very small). Similarly, the fine structure constant, $\alpha = e^2/\hbar c \simeq 1/137$, which determines the intensity of electromagnetic interactions is a function of g_s^2. Because of these relations between the physical coupling constants and g_s (and therefore the value of the dilaton; $g_s = e^\Phi$), we see that if the dilaton is massless (or in other words is long-range), its value $\Phi(x)$ at a space-time point x will depend on the distribution of matter in the universe. For example, as is the case with the gravitational field (for example $g_{00}(x) \simeq -1 + 2GM/c^2r$), we expect that the value of $\Phi(x)$ depends on the masses present around the point x, and should be different at the Earth's surface than it is at a higher altitude. One may also expect that $\Phi(x)$ would be sensitive to the expansion of the universe and would vary over a time scale comparable to the age of the universe. However, if $\Phi(x)$ varies over space and/or time, one concludes from the relations shown above between $g_s = e^\Phi$ and the physical coupling constants that the latter must also vary over space and/or time. Therefore, for example, the value, here and now, of the fine structure constant α could be slightly different from the value it had, long ago, in a very distant galaxy. Such effects are accessible to detailed astronomical observations and, in fact, some recent observations have suggested that the interaction constants were different in distant galaxies. However, other experimental data (such as the fossil nuclear reactor at Oklo and the isotopic composition of ancient terrestrial meteorites) put very severe limits on any variability of the coupling "constants". Let us finally note that if the fine structure "constant" α, as well as other coupling "constants", varies with a massless field such as the dilaton $\Phi(x)$, then this implies a violation of the basic postulate of general relativity: the principle of equivalence. In particular, one can show that the universality of free fall is necessarily violated, meaning that bodies with different nuclear composition would fall with different accelerations in an external gravitational field. This gives an important motivation for testing the principle of equivalence with greater precision. For example, the MICROSCOPE space mission [27] (of the CNES) should soon test the universality of free fall to the level of 10^{-15}, and the STEP space project (Satellite Test of the Equivalence Principle) [28] could reach the level 10^{-18}.

Another interesting phenomenological possibility is that the dilaton (and/or other scalar fields of the same type, called *moduli*) acquires a non-zero mass that is however very small with respect to the string mass scale m_s. One could then observe a modification of Newtonian gravitation over small distances (smaller than a tenth of a millimeter). For a discussion of this theoretical possibility and of its recent experimental tests see, respectively, the contributions by I. Antoniadis and J. Mester to this Poincaré seminar.

12. Conclusion

For a long time general relativity was admired as a marvelous intellectual construction, but it only played a marginal role in physics. Typical of the appraisal of this theory is the comment by Max Born [29] made upon the fiftieth anniversary of the *annus mirabilis*: "The foundations of general relativity seemed to me then, and they still do today, to be the greatest feat of human thought concerning Nature, the most astounding association of philosophical penetration, physical intuition, and mathematical ability. However its connections to experiment were tenuous. It seduced me like a great work of art that should be appreciated and admired from a distance."

Today, one century after the *annus mirabilis*, the situation is quite different. General relativity plays a central role in a large domain of physics, including everything from primordial cosmology and the physics of black holes to the observation of binary pulsars and the definition of international atomic time. It even has everyday practical applications, via the satellite positioning systems (such as the GPS and, soon, its European counterpart Galileo). Many ambitious (and costly) experimental projects aim to test it (G.P.B., MICROSCOPE, STEP, ...), or use it as a tool for deciphering the distant universe (LIGO/VIRGO/GEO, LISA, ...). The time is therefore long-gone that its connection with experiment was tenuous. Nevertheless, it is worth noting that the fascination with the structure and physical implications of the theory evoked by Born remains intact. One of the motivations for thinking that the theory of strings (and other extended objects) holds the key to the problem of the unification of physics is its deep affinity with general relativity. Indeed, while the attempts at "Grand Unification" made in the 1970s completely ignored the gravitational interaction, string theory necessarily leads to Einstein's fundamental concept of a dynamical space-time. At any rate, it seems that one must more deeply understand the "generalized quantum geometry" created through the interaction of strings and p-branes in order to completely formulate this theory and to understand its hidden symmetries and physical implications. Einstein would no doubt appreciate seeing the key role played by symmetry principles and gravity within modern physics.

References

[1] A. Einstein, *Zur Elektrodynamik bewegter Körper*, Annalen der Physik **17** (1905), 891.

[2] See http://www.einstein.caltech.edu for an entry into the Einstein Collected Papers Project. The French reader will have access to Einstein's main papers in *Albert Einstein, Œuvres choisies*, Paris, Le Seuil/CNRS, 1993, under the direction of F. Balibar. See in particular Volumes 2 (Relativités I) and 3 (Relativités II). One can also consult the 2005 Poincaré seminar dedicated to Einstein (http://www.lpthe.jussieu.fr/poincare): *Einstein, 1905–2005, Poincaré Seminar 2005*, edited by T. Damour, O. Darrigol, B. Duplantier and V. Rivasseau (Birkhäuser Verlag, Basel, Suisse, 2006). See also the excellent summary article by D. Giulini and N. Straumann, "Einstein's impact on the physics of the twentieth century," Studies in History and Philosophy of Modern Physics **37** (2006), 115–173. For online access to many of Einstein's original articles and to documents about him, see http://www.alberteinstein.info/. We also note that most of the work in progress on general relativity can be consulted on various archives at http://xxx.lanl.gov, in particular the archive gr-qc. Review articles on certain sub-fields of general relativity are accessible at http://relativity.livingreviews.org. Finally, see T. Damour, *Once Upon Einstein*, A K Peters Ltd, Wellesley, 2006, for a recent non-technical account of the formation of Einstein's ideas.

[3] Galileo, *Dialogues Concerning Two New Sciences*, translated by Henry Crew and Alfonso di Salvio, Macmillan, New York, 1914.

[4] The reader interested in learning about recent experimental tests of gravitational theories may consult, on the internet, either the highly detailed review by C.M. Will in Living Reviews (http://relativity.livingreviews.org/Articles/lrr-2001-4) or the brief review by T. Damour in the Review of Particle Physics (http://pdg.lbl.gov/). See also John Mester's contribution to this Poincaré seminar.

[5] A. Einstein, *Die Feldgleichungen der Gravitation, Sitz. Preuss. Akad. Wiss.*, 1915, p. 844.

[6] The reader wishing to study the formalism and applications of general relativity in detail can consult, for example, the following works: L. Landau and E. Lifshitz, *The Classical Theory of Fields*, Butterworth-Heinemann, 1995; S. Weinberg, *Gravitation and Cosmology*, Wiley, New York, 1972; H.C. Ohanian and R. Ruffini, *Gravitation and Spacetime*, Second Edition, Norton, New York, 1994; N. Straumann, *General Relativity, With Applications to Astrophysics*, Springer Verlag, 2004. Let us also mention detailed course notes on general relativity by S.M. Carroll, available on the internet: http://pancake.uchicago.edu/~carroll/notes/ ; as well as at gr-qc/9712019. Finally, let us mention the recent book (in French) on the history of the discovery and reception of general relativity: J. Eisenstaedt, *Einstein et la relativité générale*, CNRS, Paris, 2002.

[7] B. Bertotti, L. Iess, and P. Tortora, *A Test of General Relativity Using Radio Links with the Cassini Spacecraft*, Nature **425** (2003), 374.

[8] http://einstein.stanford.edu

[9] W. Israel, *Dark stars: the evolution of an idea*, in *300 Years of Gravitation*, edited by S.W. Hawking and W. Israel, Cambridge University Press, Cambridge, 1987, Chapter 7, pp. 199–276.

[10] The discovery of binary pulsars is related in Hulse's Nobel Lecture: R.A. Hulse, *Reviews of Modern Physics* **66** (1994), 699.

[11] For an introduction to the observational characteristics of pulsars, and their use in testing relativistic gravitation, see Taylor's Nobel Lecture: J.H. Taylor, *Reviews of Modern Physics* **66** (1994), 711. See also Michael Kramer's contribution to this Poincaré seminar.

[12] For an update on the observational characteristics of pulsars, and their use in testing general relativity, see the Living Review by I.H. Stairs, available at http://relativity.livingreviews.org/Articles/lrr-2003-5/ and the contribution by Michael Kramer to this Poincaré seminar.

[13] For a recent update on tests of relativistic gravitation (and of tensor-scalar theories) obtained through the chronometry of binary pulsars, see G. Esposito-Farèse, gr-qc/0402007 (available on the general relativity and quantum cosmology archive at the address http://xxx.lanl.gov), and T. Damour and G. Esposito-Farèse, in preparation. Figure 4 is adapted from these references.

[14] For a review of the problem of the motion of two gravitationally condensed bodies in general relativity, up to the level where the effects connected to the finite speed of propagation of the gravitational interaction appear, see T. Damour, *The problem of motion in Newtonian and Einsteinian gravity*, in *300 Years of Gravitation*, edited by S.W. Hawking and W. Israel, Cambridge University Press, Cambridge, 1987, Chapter 6, pp. 128–198.

[15] A. Einstein, *Näherungsweise Integration der Feldgleichungen der Gravitation, Sitz. Preuss. Akad. Wiss.* (1916), p. 688 ; *ibidem*, *Über Gravitationswellen* (1918), p. 154.

[16] For a highly detailed introduction to these three problems, see K.S. Thorne *Gravitational radiation*, in *300 Years of Gravitation*, edited by S.W. Hawking and W. Israel, Cambridge University Press, Cambridge, 1987, Chapter 9, pp. 330–458.

[17] http://www.ligo.caltech.edu/

[18] http://www.virgo.infn.it/

[19] http://www/geo600.uni-hanover.de/

[20] http://lisa.jpl.nasa.gov/

[21] L. Blanchet et al., gr-qc/0406012 ; see also the Living Review by L. Blanchet, available at http://relativity.livingreviews.org/Articles.

[22] Figure 5 is adapted from work by A. Buonanno and T. Damour, gr-qc/0001013.

[23] F. Pretorius, Phys. Rev. Lett. **95** (2005), 121101, gr-qc/0507014; M. Campanelli et al., Phys. Rev. Lett. **96** (2006), 111101, gr-qc/0511048 J. Baker et al., Phys. Rev. D **73** (2006), 104002, gr-qc/0602026.

[24] For a particularly clear exposé of the development of the quantum theory of fields, see, for example, the first chapter of S. Weinberg, *The Quantum Theory of Fields*, volume 1, Foundations, Cambridge University Press, Cambridge, 1995.

[25] For an introduction to the theory of (super)strings see http://superstringtheory.com/. For a detailed (and technical) introduction to the theory see the books: K. Becker, M. Becker, and J.H. Schwarz, *String Theory*

and M-theory: An Introduction, Cambridge University Press, Cambridge, 2006; B. Zwiebach, *A First Course in String Theory*, Cambridge University Press, Cambridge, 2004; M.B. Green, J.H. Schwarz et E. Witten, *Superstring theory*, 2 volumes, Cambridge University Press, Cambridge, 1987 ; and J. Polchinski, *String Theory*, 2 volumes, Cambridge University Press, Cambridge, 1998. To read review articles or to research this theory as it develops see the hep-th archive at http://xxx.lanl.gov. To search for information on the string theory literature (and more generally that of high-energy physics) see also the site http://www.slac.stanford.edu/spires/find/hep.

[26] For a detailed introduction to black hole physics see P.K. Townsend, gr-qc/9707012; for an entry into the vast literature on black hole entropy, see, for example, T. Damour, hep-th/0401160 in *Poincaré Seminar 2003*, edited by Jean Dalibard, Bertrand Duplantier, and Vincent Rivasseau (Birkhäuser Verlag, Basel, 2004), pp. 227-264.

[27] http://www.onera.fr/microscope/

[28] http://www.sstd.rl.ac.uk/fundphys/step/.

[29] M. Born, *Physics and Relativity*, in Fünfzig Jahre Relativitätstheorie, Bern, 11-16 Juli 1955, Verhandlungen, edited by A. Mercier and M. Kervaire, Helvetica Physica Acta, Supplement **4** (1956), 244–260.

Thibault Damour
Institut des Hautes Etudes Scientifiques
35, route de Chartres
91440 Bures-sur-Yvette
France
e-mail: damour@ihes.fr

Gravitation and Experiment, 51–67

Poincaré Seminar 2006

Beyond Einstein's Gravity

Ignatios Antoniadis

Abstract. Despite the important experimental success of General Relativity, there are several theoretical reasons indicating that gravitational phenomena may change radically from the predictions of Einstein's theory at very short distances. A main motivation comes from studies of unifying all fundamental forces in the framework of a consistent quantum theory, called string theory. This theory introduces a new physical constant, the string length, under which a new elementary structure shows up, changing drastically all physical laws of nature. In particular, lowering the string scale in the TeV region provides a theoretical framework for solving the so-called mass hierarchy problem: the apparent weakness of gravity can then be accounted by the existence of large internal dimensions, in the submillimeter region, and transverse to a braneworld where our observed universe is confined. I review the main properties of this scenario and its implications for new gravitational phenomena that can be observed at both particle colliders, and in non-accelerator experiments searching for new short range forces at submillimeter distances. I also discuss the warped metric case and possible localization of gravity in the presence of infinite size extra dimensions that can modify Newton's law at cosmological distance scales.

1. Introduction

During the last few decades, physics beyond the Standard Model (SM) was guided by the problem of mass hierarchy. This can be formulated as the question of why gravity appears to us so weak compared to the other three known fundamental interactions corresponding to the electromagnetic, weak and strong nuclear forces. Indeed, gravitational interactions are suppressed by a very high energy scale, the Planck mass $M_P \sim 10^{19}$ GeV, associated to a length $l_P \sim 10^{-35}$ m, where they are expected to become important. In a quantum theory, the hierarchy implies a severe fine tuning of the fundamental parameters in more than 30 decimal places in order to keep the masses of elementary particles at their observed values. The reason is that quantum radiative corrections to all masses generated by the Higgs vacuum expectation value (VEV) are proportional to the ultraviolet cutoff which in the presence of gravity is fixed by the Planck mass. As a result, all masses are "attracted" to become about 10^{16} times heavier than their observed values.

Besides compositeness, there are two main ideas that have been proposed and studied extensively during the last years, corresponding to different approaches of explaining the mass hierarchy problem. (1) Low energy supersymmetry with all superparticle masses in the TeV region. Indeed, in the limit of exact supersymmetry, quadratically divergent corrections to the Higgs self-energy are exactly cancelled, while in the softly broken case, they are cutoff by the supersymmetry breaking mass splittings. (2) TeV scale strings, in which quadratic divergences are cutoff by the string scale and low energy supersymmetry is not needed. Both ideas are experimentally testable at high-energy particle colliders and in particular at LHC. Below, I discuss their implementation in string theory.

The appropriate and most convenient framework for low energy supersymmetry and grand unification is the perturbative heterotic string. Indeed, in this theory, gravity and gauge interactions have the same origin, as massless modes of the closed heterotic string, and they are unified at the string scale M_s. As a result, the Planck mass M_P is predicted to be proportional to M_s:

$$M_P = M_s/g\,, \tag{1}$$

where g is the gauge coupling. In the simplest constructions all gauge couplings are the same at the string scale, given by the four-dimensional (4d) string coupling, and thus no grand unified group is needed for unification. In our conventions $\alpha_{\text{GUT}} = g^2 \simeq 0.04$, leading to a discrepancy between the string and grand unification scale M_{GUT} by almost two orders of magnitude. Explaining this gap introduces in general new parameters or a new scale, and the predictive power is essentially lost. This is the main defect of this framework, which remains though an open and interesting possibility.

The other idea can be naturally realized in the framework of type I string theory with D-branes. Unlike in the heterotic string, gauge and gravitational interactions have now different origin. The latter are described again by closed strings, while the former emerge as excitations of open strings with endpoints confined on D-branes [1]. This leads to a braneworld description of our universe, which should be localized on a hypersurface, i.e. a membrane extended in p spatial dimensions, called p-brane (see Fig. 1). Closed strings propagate in all nine dimensions of string theory: in those extended along the p-brane, called parallel, as well as in the transverse ones. On the contrary, open strings are attached on the p-brane. Obviously, our p-brane world must have at least the three known dimensions of space. But it may contain more: the extra $d_\parallel = p-3$ parallel dimensions must have a finite size, in order to be unobservable at present energies, and can be as large as $\text{TeV}^{-1} \sim 10^{-18}$ m [2]. On the other hand, transverse dimensions interact with us only gravitationally and experimental bounds are much weaker: their size should be less than about 0.1 mm [3]. In the following, I review the main properties and experimental signatures of low string scale models [4, 5].

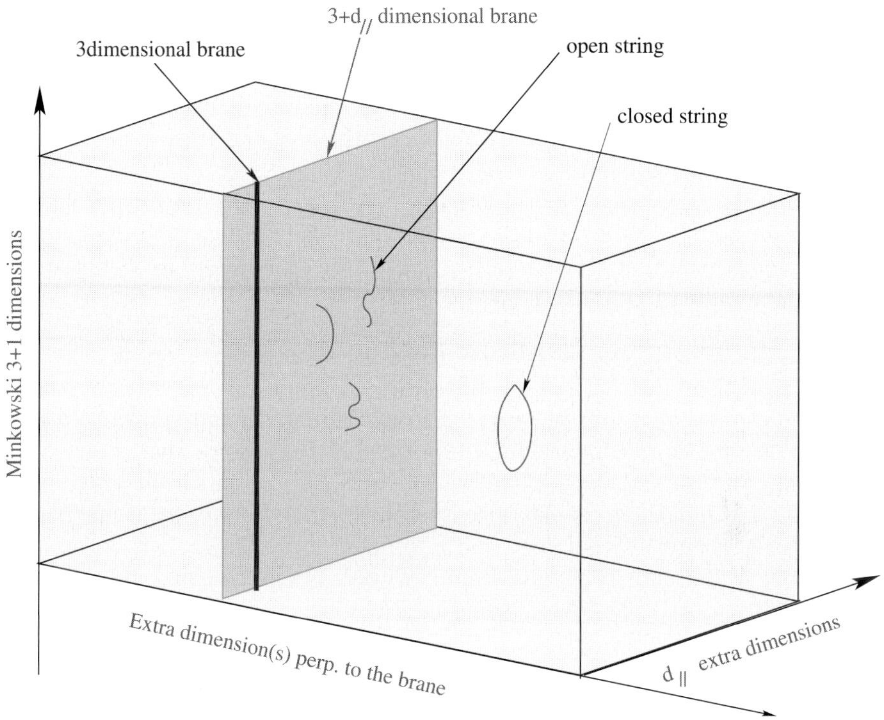

FIGURE 1. In the type I string framework, our Universe contains, besides the three known spatial dimensions (denoted by a single black line), some extra dimensions ($d_{\parallel} = p - 3$) parallel to our world p-brane (gray plane) where endpoints of open strings are confined, as well as some transverse dimensions (enclosed space) where only gravity described by closed strings can propagate.

2. Framework

In type I theory, the different origin of gauge and gravitational interactions implies that the relation between the Planck and string scales is not linear as (1) of the heterotic string. The requirement that string theory should be weakly coupled, constrain the size of all parallel dimensions to be of order of the string length, while transverse dimensions remain unrestricted. Assuming an isotropic transverse space of $n = 9 - p$ compact dimensions of common radius $R_{\perp}$, one finds:

$$M_P^2 = \frac{1}{g^4} M_s^{2+n} R_{\perp}^n \,, \qquad g_s \simeq g^2 \,. \tag{2}$$

where g_s is the string coupling. It follows that the type I string scale can be chosen hierarchically smaller than the Planck mass [6, 4] at the expense of introducing extra large transverse dimensions felt only by gravity, while keeping the string coupling small [4]. The weakness of 4d gravity compared to gauge interactions (ratio M_W/M_P) is then attributed to the largeness of the transverse space $R_\perp$ compared to the string length $l_s = M_s^{-1}$.

An important property of these models is that gravity becomes effectively $(4+n)$-dimensional with a strength comparable to those of gauge interactions at the string scale. The first relation of Eq. (2) can be understood as a consequence of the $(4+n)$-dimensional Gauss law for gravity, with

$$M_{(4+n)} = M_s/g^{4/2+n} \tag{3}$$

the effective scale of gravity in $4+n$ dimensions. Taking $M_s \simeq 1$ TeV, one finds a size for the extra dimensions $R_\perp$ varying from 10^8 km, 0.1 mm, down to a Fermi for $n = 1, 2$, or 6 large dimensions, respectively. This shows that while $n = 1$ is excluded, $n \geq 2$ is allowed by present experimental bounds on gravitational forces [3, 7]. Thus, in these models, gravity appears to us very weak at macroscopic scales because its intensity is spread in the "hidden" extra dimensions. At distances shorter than $R_\perp$, it should deviate from Newton's law, which may be possible to explore in laboratory experiments (see Fig. 2).

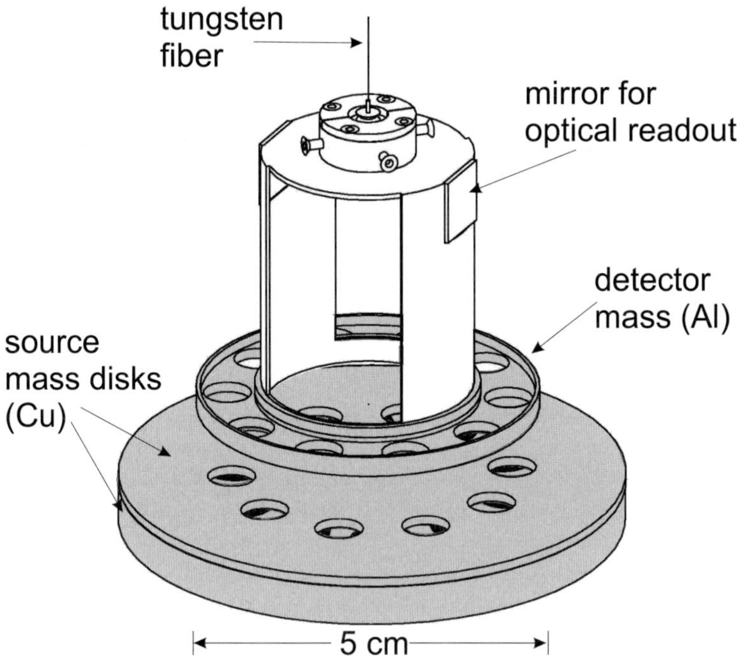

FIGURE 2. Torsion pendulum that tested Newton's law at 130 nm.

3. Experimental implications in accelerators

The main experimental signal is gravitational radiation in the bulk from any physical process on the world-brane. In fact, the very existence of branes breaks translation invariance in the transverse dimensions and gravitons can be emitted from the brane into the bulk. During a collision of center of mass energy $\sqrt{s}$, there are $\sim (\sqrt{s}R_\perp)^n$ KK excitations of gravitons with tiny masses, that can be emitted. Each of these states looks from the 4d point of view as a massive, quasi-stable, extremely weakly coupled (s/M_P^2 suppressed) particle that escapes from the detector. The total effect is a missing-energy cross-section roughly of order:

$$\frac{(\sqrt{s}R_\perp)^n}{M_P^2} \sim \frac{1}{s}\left(\frac{\sqrt{s}}{M_s}\right)^{n+2}. \tag{4}$$

Explicit computation of these effects leads to the bounds given in Table 1. However, larger radii are allowed if one relaxes the assumption of isotropy, by taking for instance two large dimensions with different radii.

TABLE 1. Limits on $R_\perp$ in mm.

Experiment	$n = 2$	$n = 4$	$n = 6$
Collider bounds			
LEP 2	5×10^{-1}	2×10^{-8}	7×10^{-11}
Tevatron	5×10^{-1}	10^{-8}	4×10^{-11}
LHC	4×10^{-3}	6×10^{-10}	3×10^{-12}
NLC	10^{-2}	10^{-9}	6×10^{-12}
Present non-collider bounds			
SN1987A	3×10^{-4}	10^{-8}	6×10^{-10}
COMPTEL	5×10^{-5}	-	-

Fig. 3 shows the cross-section for graviton emission in the bulk, corresponding to the process $pp \to jet + graviton$ at LHC, together with the SM background [8]. For a given value of M_s, the cross-section for graviton emission decreases with the number of large transverse dimensions, in contrast to the case of parallel dimensions. The reason is that gravity becomes weaker if there are more dimensions because there is more space for the gravitational field to escape. There is a particular energy and angular distribution of the produced gravitons that arise from the distribution in mass of KK states of spin-2. This can be contrasted to other sources of missing energy and might be a smoking gun for the extra dimensional nature of such a signal.

In Table 1, there are also included astrophysical and cosmological bounds. Astrophysical bounds [9, 10] arise from the requirement that the radiation of gravitons should not carry on too much of the gravitational binding energy released during core collapse of supernovae. In fact, the measurements of Kamiokande and

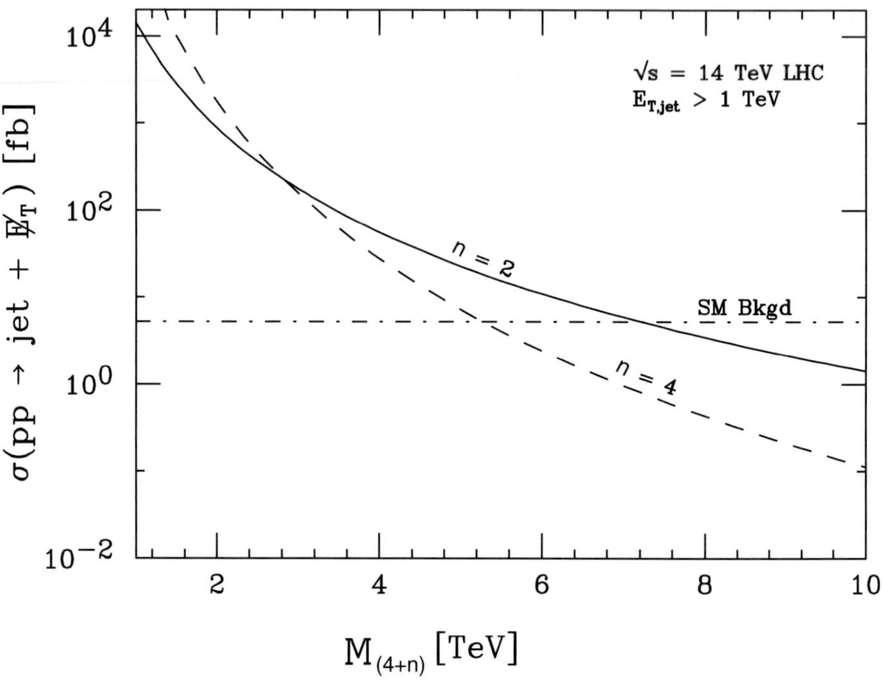

FIGURE 3. Missing energy cross-section due to graviton emission at LHC, as a function of the higher-dimensional gravity scale $M_{(4+n)}$, for n extra dimensions, produced together with a hadronic jet.

IMB for SN1987A suggest that the main channel is neutrino fluxes. The best cosmological bound [11] is obtained from requiring that decay of bulk gravitons to photons do not generate a spike in the energy spectrum of the photon background measured by the COMPTEL instrument. Bulk gravitons are expected to be produced just before nucleosynthesis due to thermal radiation from the brane. The limits assume that the temperature was at most 1 MeV as nucleosynthesis begins, and become stronger if temperature is increased.

At energies higher than the string scale, new spectacular phenomena are expected to occur, related to string physics and quantum gravity effects, such as possible micro-black hole production [12]. Particle accelerators would then become the best tools for studying quantum gravity and string theory.

4. Supersymmetry in the bulk and short range forces

Besides the spectacular predictions in accelerators, there are also modifications of gravitation in the sub-millimeter range, which can be tested in "table-top" experiments that measure gravity at short distances. There are three categories of such predictions:
(i) Deviations from the Newton's law $1/r^2$ behavior to $1/r^{2+n}$, which can be observable for $n = 2$ large transverse dimensions of sub-millimeter size. This case is particularly attractive on theoretical grounds because of the logarithmic sensitivity of SM couplings on the size of transverse space [13], that allows to determine the hierarchy [14].
(ii) New scalar forces in the sub-millimeter range, related to the mechanism of supersymmetry breaking, and mediated by light scalar fields φ with masses [15, 4]:

$$m_\varphi \simeq \frac{m_{susy}^2}{M_P} \simeq 10^{-4} - 10^{-6}\ \text{eV}\,, \tag{5}$$

for a supersymmetry breaking scale $m_{susy} \simeq 1 - 10$ TeV. They correspond to Compton wavelengths of 1 mm to 10 μm. m_{susy} can be either $1/R_\parallel$ if supersymmetry is broken by compactification [15], or the string scale if it is broken "maximally" on our world-brane [4]. A universal attractive scalar force is mediated by the radion modulus $\varphi \equiv M_P \ln R$, with R the radius of the longitudinal or transverse dimension(s). In the former case, the result (5) follows from the behavior of the vacuum energy density $\Lambda \sim 1/R_\parallel^4$ for large $R_\parallel$ (up to logarithmic corrections). In the latter, supersymmetry is broken primarily on the brane, and thus its transmission to the bulk is gravitationally suppressed, leading to (5). For $n = 2$, there may be an enhancement factor of the radion mass by $\ln R_\perp M_s \simeq 30$ decreasing its wavelength by an order of magnitude [14].

The coupling of the radius modulus to matter relative to gravity can be easily computed and is given by:

$$\sqrt{\alpha_\varphi} = \frac{1}{M}\frac{\partial M}{\partial \varphi}\ ; \quad \alpha_\varphi = \begin{cases} \frac{\partial \ln \Lambda_{QCD}}{\partial \ln R} \simeq \frac{1}{3} & \text{for } R_\parallel \\ \frac{2n}{n+2} = 1 - 1.5 & \text{for } R_\perp \end{cases} \tag{6}$$

where M denotes a generic physical mass. In the longitudinal case, the coupling arises dominantly through the radius dependence of the QCD gauge coupling [15], while in the case of transverse dimension, it can be deduced from the rescaling of the metric which changes the string to the Einstein frame and depends slightly on the bulk dimensionality ($\alpha = 1 - 1.5$ for $n = 2 - 6$) [14]. Such a force can be tested in microgravity experiments and should be contrasted with the change of Newton's law due the presence of extra dimensions that is observable only for $n = 2$ [3, 7]. The resulting bounds from an analysis of the radion effects are [3]:

$$M_* \gtrsim 3 - 4.5\,\text{TeV} \quad \text{for} \quad n = 2 - 6\,. \tag{7}$$

In principle there can be other light moduli which couple with even larger strengths. For example the dilaton, whose VEV determines the string coupling, if it does not

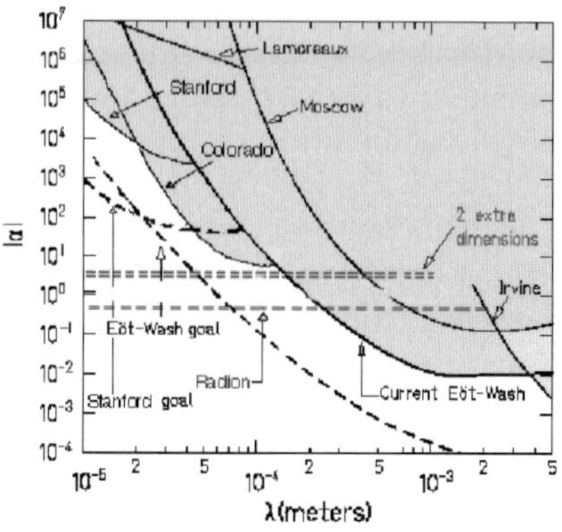

FIGURE 4. Present limits on non-Newtonian forces at short distances (gray regions), as a function of their range λ and their strength relative to gravity α. The limits are compared to new forces mediated by the graviton in the case of two large extra dimensions, and by the radion.

acquire large mass from some dynamical supersymmetric mechanism, can lead to a force of strength 2000 times bigger than gravity [16].
(iii) Non universal repulsive forces much stronger than gravity, mediated by possible abelian gauge fields in the bulk [9, 17]. Such fields acquire tiny masses of the order of M_s^2/M_P, as in (5), due to brane localized anomalies [17]. Although their gauge coupling is infinitesimally small, $g_A \sim M_s/M_P \simeq 10^{-16}$, it is still bigger than the gravitational coupling E/M_P for typical energies $E \sim 1$ GeV, and the strength of the new force would be $10^6 - 10^8$ stronger than gravity. This is an interesting region which will be soon explored in micro-gravity experiments (see Fig. 4). Note that in this case supernova constraints impose that there should be at least four large extra dimensions in the bulk [9].

In Fig. 4 we depict the actual information from previous, present and upcoming experiments [7, 14]. The solid lines indicate the present limits from the experiments indicated. The excluded regions lie above these solid lines. Measuring gravitational strength forces at short distances is challenging. The dashed thick lines give the expected sensitivity of the various experiments, which will improve the actual limits by roughly two orders of magnitude, while the horizontal dashed lines correspond to the theoretical predictions for the graviton in the case $n = 2$ and for the radion in the transverse case. These limits are compared to those obtained from particle accelerator experiments in Table 1. Finally, in Figs. 5 and 6, we display recent improved bounds for new forces at very short distances by focusing on the right hand side of Fig. 4, near the origin [7].

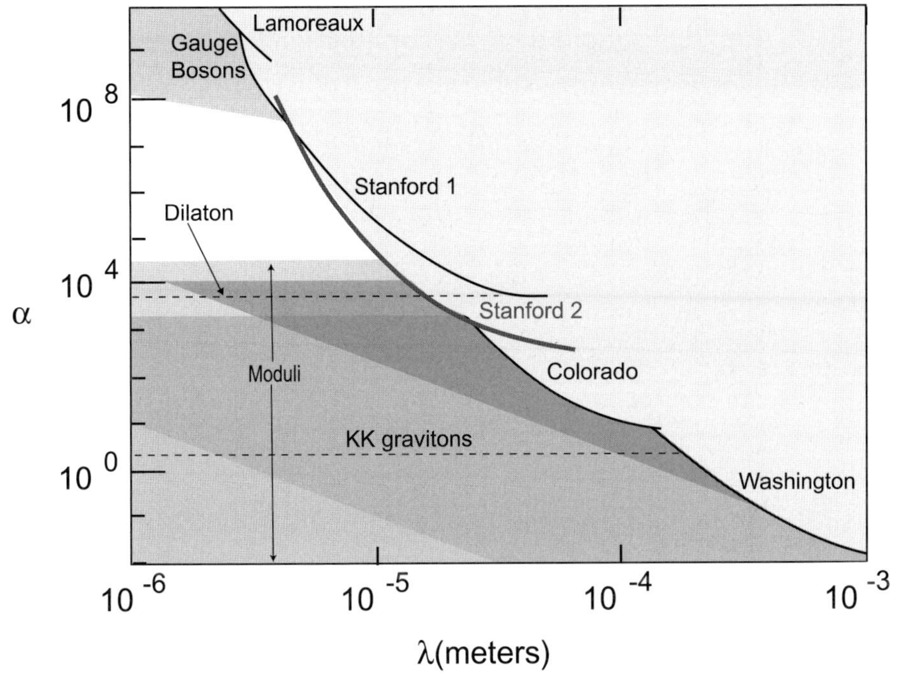

FIGURE 5. Bounds on non-Newtonian forces in the range 6-20 μm (see S. J. Smullin et al. in Ref. [7]).

5. Non-compact extra dimensions and localized gravity

There are several motivations to study localization of gravity in non-compact extra dimensions: (i) it avoids the problem of fixing the moduli associated to the size of the compactification manifold; (ii) it provides a new approach to the mass hierarchy problem; (iii) there are modifications of gravity at large distances that may have interesting observational consequences. Two types of models have been studied: warped metrics in curved space [18], and infinite size extra dimensions in flat space [19]. The former, although largely inspired by stringy developments and having used many string-theoretic techniques, have not yet a clear and calculable string theory realization [20]. In any case, since curved space is always difficult to handle in string theory, in the following we concentrate mainly on the latter, formulated in flat space with gravity localized on a subspace of the bulk. It turns out that these models of induced gravity have an interesting string theory realization [21] that we describe below, after presenting first a brief overview of the warped case [22].

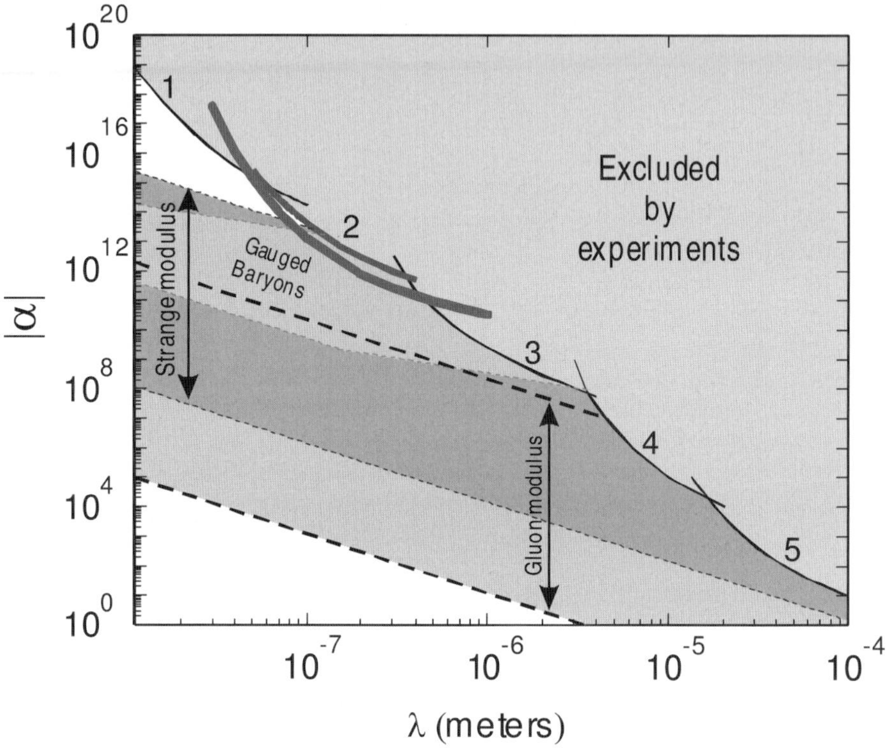

FIGURE 6. Bounds on non-Newtonian forces in the range around 200 nm (see R. S. Decca et al. in Ref. [7]). Curves 4 and 5 correspond to Stanford and Colorado experiments, respectively, of Fig. 5 (see also J C. Long and J. C. Price of Ref. [7]).

5.1. Warped spaces

In these models, space-time is a slice of anti de Sitter space (AdS) in $d = 5$ dimensions while our universe forms a four-dimensional (4d) flat boundary [18]. The corresponding line element is:

$$ds^2 = e^{-2k|y|}\eta_{\mu\nu}dx^\mu dx^\nu + dy^2 \quad ; \quad \Lambda = -24M^3k^2\,, \tag{8}$$

where M, Λ are the 5d Planck mass and cosmological constant, respectively, and the parameter k is the curvature of AdS_5. The fifth coordinate y is restricted on the interval $[0, \pi r_c]$. Thus, this model requires two 'branes', a UV and an IR, located at the two end-points of the interval, $y = 0$ and $y = \pi r_c$, respectively. The vanishing of the 4d cosmological constant requires to fine tune the two tensions: $T = -T' = 24M^3k^2$. The 4d Planck mass is given by:

$$M_P^2 = \frac{1}{k}(1 - e^{-2\pi k r_c})M^3\,. \tag{9}$$

Note that the IR brane can move to infinity by taking the limit $r_c \to \infty$, while M_P is kept finite and thus 4d gravity is always present on the brane. The reason is that the internal volume remains finite in the non-compact limit along the positive y axis. As a result, gravity is kept localized on the UV brane, while the Newtonian potential gets corrections, $1/r + 1/k^2r^3$, which are identical with those arising in the compact case of two flat extra dimensions. Using the experimental limit $k^{-1} \lesssim 0.1$ mm and the relation (9), one finds a bound for the 5d gravity scale $M \gtrsim 10^8$ GeV, corresponding to a brane tension $T \gtrsim 1$ TeV. Notice that this bound is not valid in the compact case of six extra dimensions, because their size is in the fermi range and thus the $1/r^3$ deviations of Newton's law are cut off at shorter distances.

5.2. The induced gravity model

The DGP model and its generalizations are specified by a bulk Einstein-Hilbert (EH) term and a four-dimensional EH term [19]:

$$M^{2+n} \int_{\mathcal{M}_{4+n}} d^4x d^n y \sqrt{G}\, \mathcal{R}_{(4+n)} + M_P^2 \int_{\mathcal{M}_4} d^4x \sqrt{g}\, \mathcal{R}_{(4)} \,; \quad M_P^2 \equiv r_c^n M^{2+n} \tag{10}$$

with M and M_P the (possibly independent) respective Planck scales. The scale $M \geq 1$ TeV would be related to the short-distance scale below which UV quantum gravity or stringy effects are important. The four-dimensional metric is the restriction of the bulk metric $g_{\mu\nu} = G_{\mu\nu}|$ and we assume the WORLD[1] rigid, allowing the gauge $G_{i\mu}| = 0$ with $i \geq 5$. Finally, only intrinsic curvature terms are omitted but no Gibbons–Hawking term is needed.

5.2.1. Co-dimension one. In the case of co-dimension one bulk ($n = 1$) and δ-function localization, it is easy to see that r_c is a crossover scale where gravity changes behavior on the WORLD. Indeed, by Fourier transform the quadratic part of the action (10) with respect to the 4d position x, at the WORLD position $y = 0$, one obtains $M^{2+n}(p^{2-n} + r_c^n p^2)$, where p is the 4d momentum. It follows that for distances smaller than r_c (large momenta), the first term becomes irrelevant and the graviton propagator on the "brane" exhibits four-dimensional behavior $(1/p^2)$ with Planck constant $M_P = M^3 r_c$. On the contrary, at large distances, the first term becomes dominant and the graviton propagator acquires a five-dimensional fall-off $(1/p)$ with Planck constant M. Imposing r_c to be larger than the size of the universe, $r_c \gtrsim 10^{28}$ cm, one finds $M \lesssim 100$ MeV, which seems to be in conflict with experimental bounds. However, there were arguments that these bounds can be evaded, even for values of the fundamental scale $M^{-1} \sim 1$ mm that one may need for suppressing the quantum corrections of the cosmological constant [19].

On the other hand, in the presence of non-zero brane thickness w, a new crossover length-scale seems to appear, $R_c \sim (wr_c)^{1/2}$ [23] or $r_c^{3/5} w^{2/5}$ [24].

[1] We avoid calling $\mathcal{M}_4$ a brane because, as we will see below, gravity localizes on singularities of the internal manifold, such as orbifold fixed points. Branes with localized matter can be introduced independent of gravity localization.

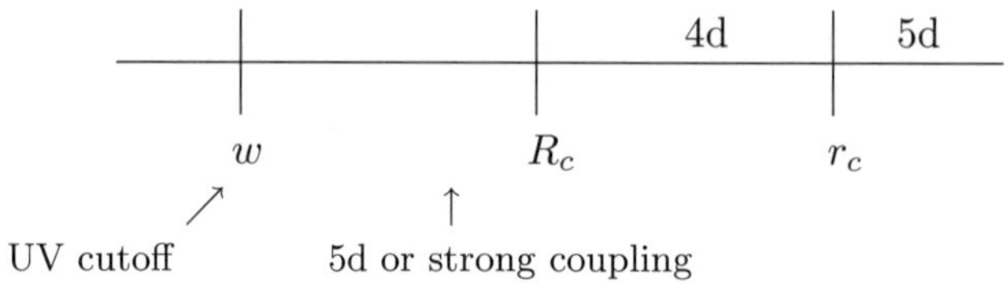

Below this scale, the theory acquires either again a five-dimensional behavior, or a strong coupling regime. For $r_c \sim 10^{28}$ cm, the new crossover scale is of order $R_c \sim 10^{-4} - 10$ m.

5.2.2. Higher co-dimension. The situation changes drastically for more than one non-compact bulk dimension, $n > 1$, due to the ultraviolet properties of the higher-dimensional theories. Indeed, from the action (10), the effective potential between two test masses in four dimensions

$$\int [d^3x]\, e^{-ip\cdot x}\, V(x) \;=\; \frac{D(p)}{1 + r_c^n\, p^2\, D(p)} \left[\tilde{T}_{\mu\nu} T^{\mu\nu} - \frac{1}{2+n}\, \tilde{T}^\mu_\mu\, T^\nu_\nu\right] \tag{11}$$

$$D(p) \;=\; \int [d^n q]\, \frac{f_w(q)}{p^2 + q^2} \tag{12}$$

is a function of the bulk graviton retarded Green's function $G(x, 0; 0, 0) = \int [d^4p]$ $e^{ip\cdot x} x D(p)$ evaluated for two points localized on the WORLD ($y = y' = 0$). The integral (12) is UV-divergent for $n > 1$ unless there is a non-trivial brane thickness profile $f_w(q)$ of width w. If the four-dimensional WORLD has zero thickness, $f_w(q) \sim 1$, the bulk graviton does not have a normalizable wave function. It therefore cannot contribute to the induced potential, which always takes the form $V(p) \sim 1/p^2$ and Newton's law remains four-dimensional at all distances.

For a non-zero thickness w, there is only one crossover length scale, R_c:

$$R_c = w \left(\frac{r_c}{w}\right)^{\frac{n}{2}}, \tag{13}$$

above which one obtains a higher-dimensional behaviour [25]. Therefore the effective potential presents two regimes: (i) at short distances ($w \ll r \ll R_c$) the gravitational interactions are mediated by the localized four-dimensional graviton and Newton's potential on the WORLD is given by $V(r) \sim 1/r$ and, (ii) at large distances ($r \gg R_c$) the modes of the bulk graviton dominate, changing the potential. Note that for $n = 1$ the expressions (11) and (12) are finite and unambiguously give $V(r) \sim 1/r$ for $r \gg r_c$. For a co-dimension bigger than 1, the precise behavior for large-distance interactions depends *crucially* on the UV completion of the theory.

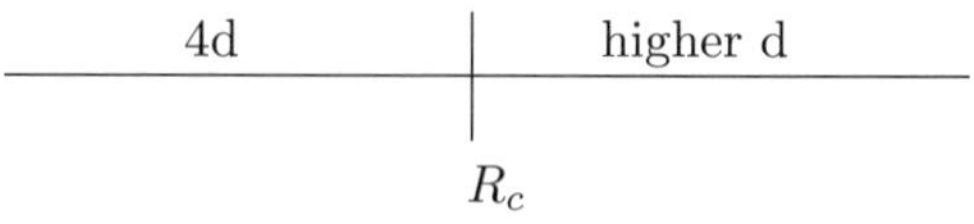

At this point we stress a fundamental difference with the *finite extra dimensions* scenarios. In these cases Newton's law gets higher-dimensional at distances smaller than the characteristic size of the extra dimensions. This is precisely the opposite of the case of infinite volume extra dimensions that we discuss here.

As mentioned above, for higher co-dimension, there is an interplay between UV regularization and IR behavior of the theory. Indeed, several works in the literature raised unitarity [26] and strong coupling problems [27] which depend crucially on the UV completion of the theory. A unitary UV regularization for the higher co-dimension version of the model has been proposed in [28]. It would be interesting to address these questions in a precise string theory context. Actually, using for UV cutoff on the "brane" the 4d Planck length $w \sim l_P$, one gets for the crossover scale (13): $R_c \sim M^{-1}(M_P/M)^{n/2}$. Putting $M \gtrsim 1$ TeV leads to $R_c \lesssim 10^{8(n-2)}$ cm. Imposing $R_c \gtrsim 10^{28}$ cm, one then finds that the number of extra dimensions must be at least six, $n \geq 6$, which is realized nicely in string theory and provides an additional motivation for studying possible string theory realizations.

5.3. String theory realization

In the following, we explain how to realize the gravity induced model (10) with $n \geq 6$ as the low-energy effective action of string theory on a non-compact six-dimensional manifold $\mathcal{M}_6$ [21]. We work in the context of $\mathcal{N} = 2$ supergravities in four dimensions but the mechanism for localizing gravity is independent of the number of supersymmetries. Of course for $\mathcal{N} \geq 3$ supersymmetries, there is no localization. We also start with the compact case and take the decompactification limit. The localized properties are then encoded in the different volume dependences.

In string perturbation, corrections to the four-dimensional Planck mass are in general very restrictive. In the heterotic string, they vanish to all orders in perturbation theory [29]; in type I theory, there are moduli-dependent corrections generated by open strings [30], but they vanish when the manifold $\mathcal{M}_6$ is decompactified; in type II theories, they are constant, independent of the moduli of the manifold $\mathcal{M}_6$, and receive contributions only from tree and one-loop levels that we describe below (at least for supersymmetric backgrounds) [21, 31]. Finally, in the context of M-theory, one obtains a similar localized action of gravity kinetic terms in five dimensions, corresponding to the strong coupling limit of type IIA string [21].

The origin of the two EH terms in (10) can be traced back to the perturbative corrections to the eight-derivative effective action of type II strings in ten dimensions. These corrections include the tree-level and one-loop terms given by:

$$\frac{1}{l_s^8}\int_{M_{10}} \frac{1}{g_s^2}\mathcal{R}_{(10)} - \frac{1}{l_s^2}\int_{M_{10}} \left(\frac{2\zeta(3)}{g_s^2} \mp 4\zeta(2)\right) R\wedge R\wedge R\wedge R\wedge e\wedge e + \cdots \qquad (14)$$

where ϕ is the dilaton field determining the string coupling $g_s = e^{\langle\phi\rangle}$, and the $\pm$ sign corresponds to the type IIA/B theory. On a direct product space-time $\mathcal{M}_6 \times \mathbb{R}^4$, at the level of zero modes, the second term in (14) splits as:

$$\int_{M_6} R\wedge R\wedge R \times \int_{M_4} \mathcal{R}_{(4)} = \chi \int_{M_4} \mathcal{R}_{(4)}\,, \tag{15}$$

where χ is the Euler number of the M_6 compactification manifold. We thus obtain the expressions for the Planck masses M and M_p:

$$M^2 \sim M_s^2/g_s^{1/2} \quad ; \quad M_P^2 \sim \chi(\frac{c_0}{g_s^2} + c_1)M_s^2\,, \tag{16}$$

with $c_0 = -2\zeta(3)$ and $c_1 = \pm 4\zeta(2) = \pm 2\pi^2/3$.

It is interesting that the appearance of the induced 4d localized term preserves $\mathcal{N}=2$ supersymmetry and is independent of the localization mechanism of matter fields (for instance on D-branes). Localization requires the internal space M_6 to have a non-zero Euler characteristic $\chi \neq 0$. Actually, in type IIA/B compactified on a Calabi-Yau manifold, χ counts the difference between the numbers of $\mathcal{N}=2$ vector multiplets and hypermultiplets: $\chi = \pm 4(n_V - n_H)$ (where the graviton multiplet counts as one vector). Moreover, in the non-compact limit, the Euler number can in general split in different singular points of the internal space, $\chi = \sum_I \chi_I$, giving rise to different localized terms at various points y_I of the internal space. A number of conclusions (confirmed by string calculations in [21]) can be reached by looking closely at (14)-(16):

▷ $M_p \gg M$ requires a large non-zero Euler characteristic for M_6, and/or a weak string coupling constant $g_s \to 0$.

▷ Since χ is a topological invariant the localized $\mathcal{R}_{(4)}$ term coming from the closed string sector is universal, independent of the background geometry and dependent only on the internal topology. It is a matter of simple inspection to see that if one wants to have a localized EH term in less than ten dimensions, namely something linear in curvature, with non-compact internal space in all directions, *the only possible dimension is four* (or five in the strong coupling M-theory limit).

▷ In order to find the width w of the localized term, one has to do a separate analysis. On general grounds, using dimensional analysis in the limit $M_P \to \infty$, one expects the effective width to vanish as a power of $l_P \equiv M_P^{-1}$: $w \sim l_P^{\nu}/l_s^{\nu-1}$ with $\nu > 0$. The computation of ν for a general Calabi-Yau space, besides its technical difficulty, presents an additional important complication: from the expression (16), $l_P \sim g_s l_s$ in the weak coupling limit. Thus, w vanishes in perturbation theory and one has to perform a non-perturbative analysis to extract its behavior. Alternatively, one can examine the case of orbifolds. In this limit, $c_0 = 0$, $l_P \sim l_s$, and the hierarchy $M_P > M$ is achieved only in the limit of large χ. One then finds that the width is given by the four-dimensional induced Planck mass

$$w \simeq l_P = l_s\,\chi^{-1/2}\,, \tag{17}$$

and the power $\nu = 1$.

5.3.1. Summary of the results. Using $w \sim l_P$ and the relations (16) in the weak coupling limit (with $c_0 \neq 0$), the crossover radius of eq. (13) is given by the string parameters ($n = 6$)

$$R_c = \frac{r_c^3}{w^2} \sim g_s \frac{l_s^4}{l_P^3} \simeq g_s \times 10^{32} \text{ cm}, \tag{18}$$

for $M_s \simeq 1$ TeV. Because R_c has to be of cosmological size, the string coupling can be relatively small, and the Euler number $|\chi| \simeq g_s^2 \, l_P \sim g_s^2 \times 10^{32}$ must be very large. The hierarchy is obtained mainly thanks to the large value of χ, so that lowering the bound on R_c lowers the value of χ. Our actual knowledge of gravity at very large distances indicates [32] that R_c should be of the order of the Hubble radius $R_c \simeq 10^{28}$ cm, which implies $g_s \geq 10^{-4}$ and $|\chi| \gtrsim 10^{24}$. A large Euler number implies only a large number of closed string massless particles with no a-priori constraint on the observable gauge and matter sectors, which can be introduced for instance on D3-branes placed at the position where gravity localization occurs. All these particles are localized at the orbifold fixed points (or where the Euler number is concentrated in the general case), and should have sufficiently suppressed gravitational-type couplings, so that their presence with such a huge multiplicity does not contradict observations. Note that these results depend crucially on the scaling of the width w in terms of the Planck length: $w \sim l_P^\nu$, implying $R_c \sim 1/l_P^{2\nu+1}$ in string units. If there are models with $\nu > 1$, the required value of χ will be much lower, becoming $\mathcal{O}(1)$ for $\nu \geq 3/2$. In this case, the hierarchy could be determined by tuning the string coupling to infinitesimal values, $g_s \sim 10^{-16}$.

The explicit string realization of localized induced gravity models offers a consistent framework that allows to address a certain number of interesting physics problems. In particular, the effective UV cutoff and the study of the gravity force among matter sources localized on D-branes. It would be also interesting to perform explicit model building and study in detail the phenomenological consequences of these models and compare to other realizations of TeV strings with compact dimensions.

Acknowledgments. This work was supported in part by the European Commission under the RTN contract MRTN-CT-2004-503369, and in part by the INTAS contract 03-51-6346.

References

[1] C. Angelantonj and A. Sagnotti, Phys. Rept. **371** (2002), 1 [Erratum-ibid. **376** (2003), 339] [arXiv:hep-th/0204089].

[2] I. Antoniadis, Phys. Lett. B **246** (1990), 377.

[3] C. D. Hoyle, D. J. Kapner, B. R. Heckel, E. G. Adelberger, J. H. Gundlach, U. Schmidt and H. E. Swanson, Phys. Rev. D **70** (2004), 042004.

[4] N. Arkani-Hamed, S. Dimopoulos and G. R. Dvali, Phys. Lett. B **429** (1998), 263 [arXiv:hep-ph/9803315]; I. Antoniadis, N. Arkani-Hamed, S. Dimopoulos and G. R. Dvali, Phys. Lett. B **436** (1998), 257 [arXiv:hep-ph/9804398].

[5] For a review see, e.g. I. Antoniadis, *Prepared for NATO Advanced Study Institute and EC Summer School on Progress in String, Field and Particle Theory, Cargese, Corsica, France (2002)*; and references therein.

[6] J. D. Lykken, Phys. Rev. D **54** (1996), 3693 [arXiv:hep-th/9603133].

[7] J. C. Long and J. C. Price, *Comptes Rendus Physique* **4** (2003), 337; R. S. Decca, D. Lopez, H. B. Chan, E. Fischbach, D. E. Krause and C. R. Jamell, Phys. Rev. Lett. **94** (2005), 240401; S. J. Smullin, A. A. Geraci, D. M. Weld, J. Chiaverini, S. Holmes and A. Kapitulnik, arXiv:hep-ph/0508204; H. Abele, S. Haeßler and A. Westphal, in 271th WE-Heraeus-Seminar, Bad Honnef (2002).

[8] G. F. Giudice, R. Rattazzi and J. D. Wells, Nucl. Phys. B **544** (1999), 3 ; E. A. Mirabelli, M. Perelstein and M. E. Peskin, Phys. Rev. Lett. **82** (1999), 2236 ; T. Han, J. D. Lykken and R. Zhang, Phys. Rev. D **59** (1999), 105006 ; K. Cheung and W.-Y. Keung, Phys. Rev. D **60** (1999), 112003 ; C. Balázs *et al.*, Phys. Rev. Lett. **83** (1999), 2112 ; L3 Collaboration (M. Acciarri *et al.*), Phys. Lett. B **464** (1999), 135 and **470** (1999), 281: J. L. Hewett, Phys. Rev. Lett. **82** (1999), 4765 .

[9] N. Arkani-Hamed, S. Dimopoulos and G. Dvali, Phys. Rev. D **59** (1999), 086004 .

[10] S. Cullen and M. Perelstein, Phys. Rev. Lett. **83** (1999), 268 ; V. Barger, T. Han, C. Kao and R. J. Zhang, Phys. Lett. B **461** (1999), 34 .

[11] K. Benakli and S. Davidson, Phys. Rev. D **60** (1999), 025004 ; L. J. Hall and D. Smith, Phys. Rev. D **60** (1999), 085008 .

[12] S. B. Giddings and S. Thomas, Phys. Rev. D **65** (2002), 056010 ; S. Dimopoulos and G. Landsberg, Phys. Rev. Lett. **87** (2001), 161602.

[13] I. Antoniadis, C. Bachas, Phys. Lett. B **450** (1999), 83 .

[14] I. Antoniadis, K. Benakli, A. Laugier and T. Maillard, Nucl. Phys. B **662** (2003), 40 [arXiv:hep-ph/0211409].

[15] I. Antoniadis, S. Dimopoulos and G. Dvali, Nucl. Phys. B **516** (1998), 70 ; S. Ferrara, C. Kounnas and F. Zwirner, Nucl. Phys. B **429** (1994), 589 .

[16] T. R. Taylor and G. Veneziano, Phys. Lett. B **213** (1988), 450 .

[17] I. Antoniadis, E. Kiritsis and J. Rizos, Nucl. Phys. B **637** (2002), 92 .

[18] L. Randall and R. Sundrum, Phys. Rev. Lett. **83** (1999), 4690 and Phys. Rev. Lett. **83** (1999), 3370 .

[19] G. R. Dvali, G. Gabadadze and M. Porrati, Phys. Lett. B **485** (2000), 208 .

[20] H. Verlinde, Nucl. Phys. B **580** (2000), 264 ; S. B. Giddings, S. Kachru and J. Polchinski, Phys. Rev. D **66** (2002), 106006 .

[21] I. Antoniadis, R. Minasian and P. Vanhove, Nucl. Phys. B **648** (2003), 69 [arXiv:hep-th/0209030].

[22] For a recent review see e.g. R. Maartens, Living Rev. Rel. **7** (2004), 7 [arXiv:gr-qc/0312059]; same proceedings and references therein.

[23] E. Kiritsis, N. Tetradis and T. N. Tomaras, JHEP **0108** (2001), 012 .

[24] M. A. Luty, M. Porrati and R. Rattazzi, arXiv:hep-th/0303116.

[25] G. R. Dvali and G. Gabadadze, Phys. Rev. D **63** (2001), 065007 ; G. R. Dvali, G. Gabadadze, M. Kolanovic and F. Nitti, Phys. Rev. D **64** (2001), 084004 .

[26] S. L. Dubovsky and V. A. Rubakov, Phys. Rev. D **67** (2003), 104014 [arXiv:hep-th/0212222].

[27] V. A. Rubakov, arXiv:hep-th/0303125.

[28] M. Kolanovic, M. Porrati and J. W. Rombouts, Phys. Rev. D **68** (2003), 064018 [arXiv:hep-th/0304148].

[29] I. Antoniadis, E. Gava and K. S. Narain, Phys. Lett. B **283** (1992), 209 .

[30] I. Antoniadis, C. Bachas, C. Fabre, H. Partouche and T. R. Taylor, Nucl. Phys. B **489** (1997), 160 ; I. Antoniadis, H. Partouche and T. R. Taylor, Nucl. Phys. B **499** (1997), 29 .

[31] I. Antoniadis, S. Ferrara, R. Minasian and K. S. Narain, Nucl. Phys. B **507** (1997), 571 .

[32] A. Lue and G. Starkman, Phys. Rev. D **67** (2003), 064002 .

Ignatios Antoniadis[2]
Department of Physics
CERN - Theory Division
1211 Geneva 23, Switzerland
e-mail: `ignatios.antoniadis@cern.ch`

[2]On leave from CPHT (UMR CNRS 7644) Ecole Polytechnique, F-91128 Palaiseau

Gravitation and Experiment, 69–100

Poincaré Seminar 2006

The Double Pulsar

Michael Kramer

Abstract. A new era in fundamental physics began with the discovery of pulsars 1967, the discovery of the first binary pulsar in 1974 and the first millisecond pulsar in 1982. Ever since, pulsars have been used as precise cosmic clocks, taking us beyond the weak-field regime of the solar-system in the study of theories of gravity. Their contribution is crucial as no test can be considered to be complete without probing the strong-field realm of gravitational physics by finding and timing pulsars. This is particularly highlighted by the discovery of the first double pulsar system which was discovered by our team in 2003. The double pulsar is unique in that both neutron stars are detectable as radio pulsars. This, combined with significantly higher mean orbital velocities and accelerations when compared to other binary pulsars, suggested that the system would become the best available testbed for general relativity and alternative theories of gravity in the strong-field regime. Indeed, this has been achieved only three years after its discovery with four independent strong-field tests of GR, more than has been obtained for any other system. Use of the theory-independent mass ratio of the two stars makes these tests uniquely different from all preceding studies. Our results confirm the validity of GR at the 0.05% level, which is by far the best precision yet achieved for the strong-field regime. Remarkably, the transverse velocity of the systems center of mass is extremely small, a result which is important for future GR tests and evolutionary studies.

1. Introduction

Rarely had the formulation of a single theory changed our view of the Universe so dramatically as Einstein's theory of general relativity (GR). Immediately after its publication, scientists were considering ways of testing GR to experimentally verify the revolutionary different effects that were predicted as deviations from Newton's theory of gravity which had ruled supreme for about three hundred years. Already during World War I., British scientists were planning two expeditions to test GR by observing a predicted bending of light around the Sun during a total solar eclipse to be observed in 1919 in Brazil and Africa. The expeditions indeed took place, and Sir Arthur Eddington himself analysed photographic plates taken during the eclipse that showed the apparent displacement of stars when they were positioned behind the Sun. This bending of light near massive bodies confirmed one of Einstein's predictions and propelled him to immediate global stardom. Today, 85 years after these historic observations, physicists are still trying to put GR to the test.

To date the theory of general relativity has passed all observational tests with flying colours. Nevertheless, GR may indeed not be the last word in our understanding of gravitational physics, and it is important to experimentally confront the theory with new observations to explore different aspects and/or achieve higher precision. While some of the most stringent tests of GR are obtained by satellite experiments in the solar system, these solar-system experiments are all made in the weak-field gravitational regime. Tests of the strong-field limit, in particular involving the radiative aspects of GR, need also to be tested. For instance, it is possible to construct theories, which would pass all solar-system tests but would show deviations from GR in the strong-field limit (see, e.g., [1]). Precision tests in the strong-field regime are best achieved by observing radio pulsars.

One of the toughest tests ever has recently become possible by the discovery of two pulsars which are found to be in a close, slowly decaying orbit that will eventually lead to the destruction of these objects, predicted to happen in about 85 million years. This system is widely known simply as the *double pulsar*.

2. Pulsars

Pulsars are highly magnetised, rotating neutron stars which emit a narrow radio beam along the magnetic dipole axis. As the magnetic axis is inclined to the rotation axis, the pulsar acts like a cosmic light-house emitting a radio pulse that can be detected once per rotation period when the beam is directed toward Earth. For some very fast rotating pulsars, the so-called millisecond pulsars (see Sect. 5), the stability of the pulse period is similar to that achieved by the best terrestrial atomic clocks. Using these astrophysical clocks by accurately measuring the arrival times of their pulses, a wide range of experiments is possible. For most of these it is not necessarily important *how* the radio pulses are actually created. We will consider some of the basic pulsar properties below.

Pulsars are born in supernova explosions of massive stars. Created in the collapse of the stars' core, neutron stars are the most compact objects next to black holes. From timing measurements of binary pulsars (see Sect. 7), we determine the masses of pulsars to be typically around $1.35 \pm 0.04 M_\odot$ [2] although this range has been expanded recently from $\sim 1.2 M_\odot$ to $2.1 M_\odot$. Modern calculations for different equations of state produce results for the size of a neutron star which are quite similar to the very first calculations by Oppenheimer & Volkov [3], i.e. a diameter of about 20 km. Such sizes are consistent with independent estimates derived from X-ray light-curves and luminosities of pulsars (e.g., [4]).

Pulsars emit electromagnetic radiation and, in particular, magnetic dipole radiation as they essentially represent rotating magnets. Assuming that this is the dominant process of loss in rotational energy and hence responsible for the observed increase in rotation period, P, described by $\dot{P}$, we can equate the corresponding energy output of the dipole to the loss rate in rotational energy. We obtain an estimate for the magnetic field strength at the pulsar (equatorial) surface from

$$B_S = 3.2 \times 10^{19} \sqrt{P\dot{P}} \text{ Gauss}, \tag{1}$$

with P measured in s and $\dot{P}$ in s s^{-1}. Millisecond pulsars (see Section 5) have lower field strengths on the order of 10^8 to 10^{10} Gauss which appears to be a result of their evolutionary history. These magnetic fields are consistent with values derived from X-ray spectra of neutron stars where we observe cyclotron lines [5].

3. Pulsars as radio sources

The periodic beacon sent by the pulsar clock is usually rather weak, both because the pulsar is distant and the size of the actual emission region is small. Estimates range down to a few metres, resulting in brightness temperatures of up to 10^{37} K [6]. Such values require a coherent emission mechanism which, despite almost 40 years of intensive research, is still unidentified. However, we seem to have some basic understanding, in which the magnetised rotating neutron star induces an electric quadrupole field which is strong enough to pull out charges from the stellar surface (the electrical force exceeds the gravitational force by a factor of $\sim 10^{12}$!), surrounding the pulsar with dense plasma. The magnetic field forces the plasma to co-rotate with the pulsar like a rigid body. This co-rotating *magnetosphere* can only extend up to a distance where the co-rotation velocity reaches the speed of light[1]. This distance defines the so-called light cylinder which separates the magnetic field lines into two distinct groups, i.e. *open and closed field lines.* Closed field lines are those which close within the light cylinder, while open field lines would close outside. The plasma on the closed field lines is trapped and will co-rotate with the pulsar forever. In contrast, plasma on the open field lines can reach highly relativistic velocities and can leave the magnetosphere, creating the observed radio beam at a distance of a few tens to hundreds of km above the pulsar surface (see Fig. 1).

Most pulsars are not strong enough to allow us a detection of their individual radio pulses, so that in most cases only an integrated pulse shape can be observed. If individual pulses are observable, they reflect the instantaneous plasma processes in the pulsar magnetosphere at the moment when the beam is directed towards Earth. The dynamics of these processes results in often seemingly random individual pulses, in particular when viewed with high time resolution. Despite this

[1]Strictly speaking, the Alfvén velocity will determine the co-rotational properties of the magnetosphere.

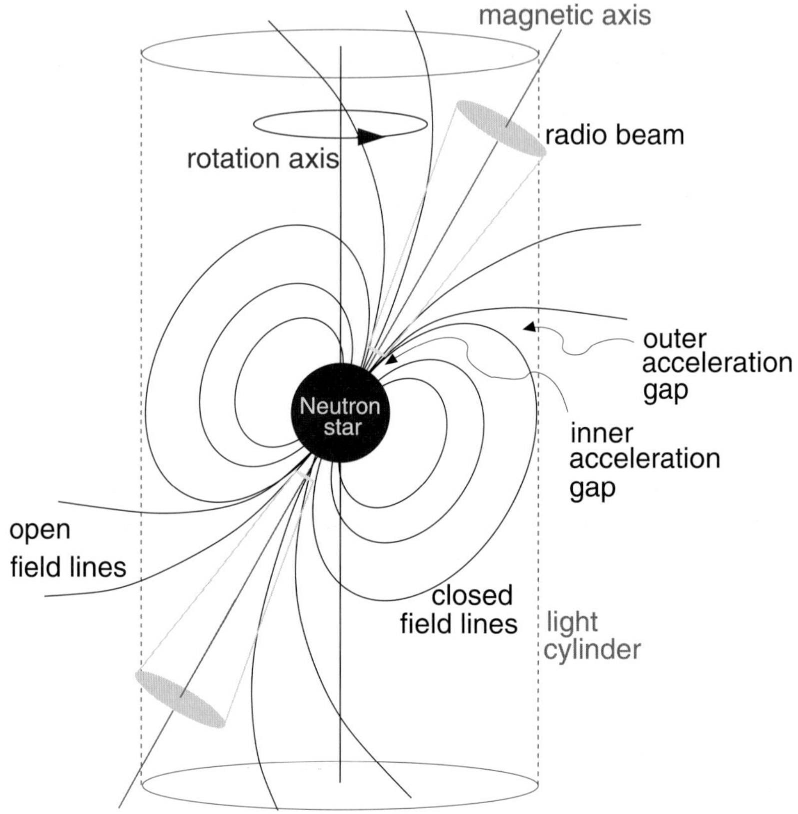

Figure 1: A pulsar is a rotating, highly magnetised neutron star. A radio beam centred on the magnetic axis is created in some distance to the pulsar. The tilt between the rotation and magnetic axes makes the pulsar in effect a cosmic lighthouse when the beam sweeps around in space.

variety displayed by the single pulses, the mean pulse shape computed by averaging a few hundreds to few thousands of pulses is incredibly stable [7]. In contrast to the snapshot provided by the individual pulses, the average pulse shape, or *pulse profile*, can be considered as a long-exposure picture, revealing the global circumstances in the magnetosphere. These are mostly determined by geometrical factors and the strong magnetic field, leading to very stable pulse profiles. Apart from a distinct evolution with radio frequency, the same profiles are obtained, no matter where and when the pulses used to compute the average have been observed.

4. Evolution of Pulsars

As pulsars are powered by their rotational energy, their spin-frequency decreases with time. The slow-down can be described by

$$\dot{\nu} = -\text{const.}\ \nu^n \tag{2}$$

where the exponent, n, is known as the *braking index*. For magnetic dipole emission as the main energy loss, we expect $n = 3$. Measuring a second spin-frequency derivative, $\ddot{\nu}$, one can obviously determine the braking index via

$$n = \nu\ddot{\nu}/\dot{\nu}^2 \tag{3}$$

so that the assumption of dipole braking can be tested. However, this is only possible for the very youngest pulsars [8], whilst rotational instabilities known as *timing noise* can mimic a significant but time-varying value of $\ddot{\nu}$. These values then reflect timing noise rather than regular spin-down, so that derived braking indices are meaningless [9]. In total, only for five young pulsars could a braking index be determined that appears to reflect the long-term spin-down behaviour. With values ranging from $n = 1.4$ to $n = 2.9$ (e.g., [10]), the deviations from the expected braking index are not too severe.

Integrating Eqn. 2, we can estimate the age of a pulsar from

$$\tau = \frac{P}{(n-1)\dot{P}} = \frac{P}{2\dot{P}} = -\frac{\nu}{2\dot{\nu}} \tag{4}$$

where we have assumed a magnetic dipole braking index of $n = 3$. This quantity known as the *characteristic age* is a valid estimate for the true age under the assumption that the initial spin period is much smaller than the present period. While it had been assumed in the past that pulsars are born with periods similar to that estimated for the Crab pulsar, $P_0 = 19$ ms [11], recent estimates suggest a wide range of initial spin periods from 14 ms up to 140 ms [12].

We can describe the evolution of a pulsar in period, P, and slow-down, $\dot{P}$, in a logarithmic P-$\dot{P}$-diagram as shown in Figure 2 where we plot all of the ~ 1700 known pulsars for which P and $\dot{P}$ have been measured. Since the estimates for both magnetic field (Eqn. 1) and characteristic age (Eqn. 4) depend only on P and $\dot{P}$, we can draw lines of constant magnetic field and constant characteristic age. Accordingly, young pulsars should be located in the upper left area of Fig. 2. Pulsars are generally considered to be young if their characteristic age is less than 100 kyr. Specifically, pulsars with characteristic ages of less than 10 kyr appear in the cross-hatched area, whilst pulsars with ages between 10 and 100 kyr are located in the hatched area. The latter pulsars are often compared to the Vela pulsar if they match its *spin-down luminosity*, i.e. $\dot{E} > 10^{36}$ erg s^{-1}. The spin-down luminosity is simply given by the loss in rotational energy which can be measured from the observed period and period derivative,

$$\dot{E} = 4\pi^2 I\dot{P}P^{-3}\ \text{erg s}^{-1} \tag{5}$$

where a neutron star moment of inertia of $I = 10^{45}$ g cm^2 is assumed. Obviously, $\dot{E}$ represents the maximum energy output available for spin-powered pulsars across the *whole* electromagnetic spectrum. A line of a constant, Vela-like $\dot{E} = 10^{36}$ erg s^{-1} is shown in Fig. 2 together with a line for $\dot{E} = 10^{33}$ erg s^{-1}.

When pulsars age, they move into the central part of the $P-\dot{P}$-diagram where they spend most of their lifetime. Consequently, most known pulsars have spin periods between 0.1 and 1.0 s with period derivatives of typically $\dot{P} = 10^{-15}$ s s^{-1}. Selection effects are only partly responsible for the limited number of pulsars known with very long periods, the longest known period being 8.5 s [13]. The dominant effect is due to the "death" of pulsars when their slow-down has reached a critical state. This state seems to depend on a combination of P and $\dot{P}$ which can be represented in the $P - \dot{P}$-diagram as a *pulsar death-line.* To the right and below this line (see Figure 2) the electric potential above the polar cap may not be sufficient to produce the particle plasma that is responsible for the observed radio emission. While this model can indeed explain the lack of pulsars beyond the death-line, the truth may be more complicated as the position of the 8.5-sec pulsar deep in the *pulsar graveyard* indicates. Nevertheless, it is clear that the normal life of radio pulsars is limited and that they die eventually after tens to a hundred million years. The position of a sub-set of about 100 pulsars located in the lower left part of Fig. 2 cannot be explained by the above picture of normal pulsar life. The evolution of these "millisecond pulsars" is different.

5. Formation of Millisecond Pulsars

The millisecond pulsars located in lower-left part of the $P - \dot{P}$ diagram are clearly different from the majority of pulsars. Firstly, they exhibit much shorter pulse periods. The first discovered millisecond pulsar, PSR B1937+21 [14], has a period of only 1.56 ms and remained the pulsar with the shortest period known for more than 20 years. Only recently, a millisecond pulsar, PSR J1748-2446ad, was discovered which has a slightly shorter period of 1.40 ms [15]. Secondly, millisecond pulsars also have very small period derivatives, $\dot{P} \lesssim 10^{-18}$ s s^{-1}, making them much older (see Eqn. 4) than normal pulsars with ages up to $\sim 10^{10}$ yr. In the standard scenario, which finds its ultimate confirmation in the discovery of the double pulsar, millisecond pulsars are recycled from a dead binary pulsar via an X-ray binary accretion phase. The pulsars' millisecond periods are obtained when mass and thereby angular momentum is transferred from an evolving binary companion while it overflows its Roche lobe (e.g., [16]).

Even though most ordinary stars are in binary systems, most pulsars do not evolve into a millisecond pulsar. The birth of the pulsar usually disrupts the system, preventing the access to a mass donor and explaining why most pulsars are isolated. In the binary systems that survive the supernova explosion, the pulsar will eventually cease to emit radio emission, before the system evolves into a X-ray binary phase during which mass accretion onto the pulsar occurs. The pulsar

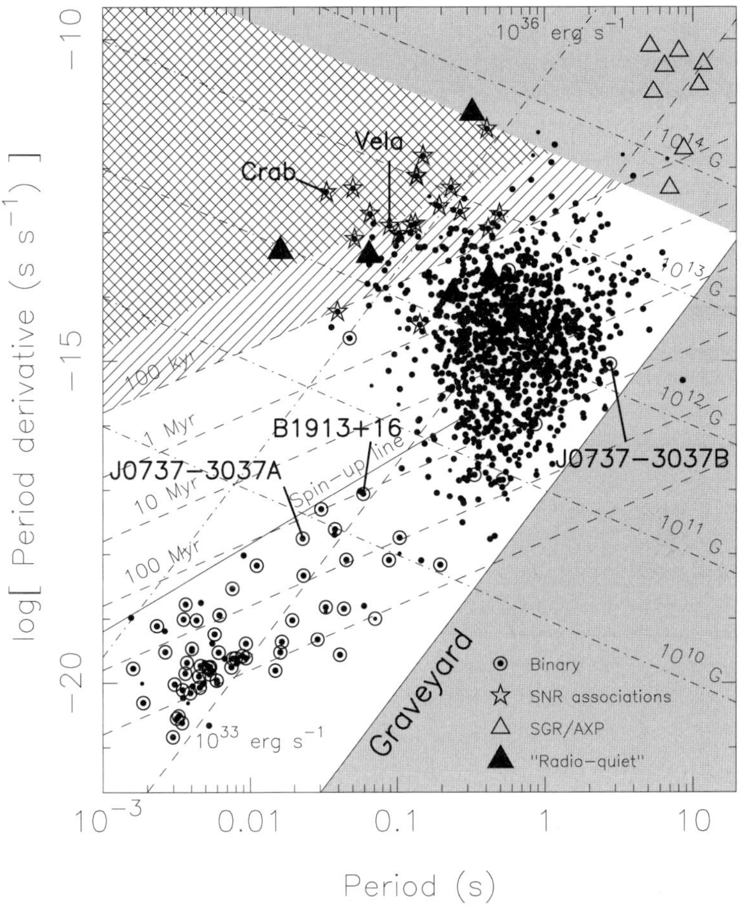

Figure 2: The $P - \dot{P}$–diagram for the known pulsar population. Lines of constant characteristic age, surface magnetic field and spin-down luminosity are shown. Binary pulsars are marked by a circle. The lower solid line represents the pulsar "death line" enclosing the "pulsar graveyard" where pulsars are expected to switch off radio emission. The grey area in the top right corner indicates the region where the surface magnetic field appears to exceed the quantum critical field of 4.4×10^{13} Gauss. For such values, some theories expect the quenching of radio emission in order to explain the radio-quiet "magnetars" (i.e. Soft-gamma ray repeaters, SGRs, and Anomalous X-ray pulsars, AXPs). The upper solid line is the "spin-up" line which is derived for the recycling process as the period limit for millisecond pulsars.

spins up and is recycled into a radio millisecond pulsar when P and $\dot{P}$ have been altered such that the pulsar has crossed the death-line again in the other direction. The final spin period of such a recycled pulsar depends on the mass of the binary companion. A more massive companion evolves faster, limiting the duration of the accretion process and hence the angular momentum transfer.

The majority of millisecond pulsars will have had a low-massive companion. These systems evolve into low-mass X-ray binaries (LMXBs) and will result into a fast-spinning millisecond pulsar with period of $P \sim 1 - 10$ ms with a low-mass white-dwarf companion. Systems with a more massive companion evolve into high-mass X-ray binaries (HMXBs) which represent the progenitors for double neutron star systems (DNSs). DNSs are rare since these systems need to survive a total of two supernova explosions. If this happens, the millisecond pulsar is only mildly recycled with a period of tens of millisecond.

The properties of millisecond pulsars and X-ray binaries are consistent with the described picture. For instance, it is striking that $\sim 80\%$ of all millisecond pulsars are in a binary orbit while this is true for only less than 1% of the non-recycled population. For millisecond pulsars with a low-mass white dwarf companion the orbit is nearly circular due to a circularisation of the orbit during the recycling process. In case of DNS systems, the orbit is affected by the unpredictable nature of the kick imparted onto the newly born neutron star in the asymmetric supernova explosion of the companion. If the system survives, the result is typically an eccentric orbit with an orbital period of a few hours.

6. Pulsar Timing

The clock-like stability of pulsars means that through precise monitoring of pulsar rotations we can study a rich variety of phenomena that affect the propagation of their pulses. While the basic spin and astrometric parameters can be derived for essentially all pulsars, millisecond pulsars are the most useful objects for more exotic applications. Their pulse arrival times can be measured much more precisely than for normal pulsars (scaling essentially with the pulse period) and their rotation is also much smoother, making them intrinsically better clocks. Specifically, they usually do not exhibit rotational instabilities such as 'timing noise' and 'glitches' known for normal pulsars.

The key quantity of interest is the *time of arrival* (TOA) of pulses at the telescope. However, since individual pulses are usually too weak to be detected, and since they also show a jitter in arrival time within a window given by the extend of the pulse profile, it is the latter which is used for timing. The stability of pulse profiles allows us to compare the observed profile with a high signal-to-noise ratio template that is constructed from previous observations. The time-offset between template and profile determines the TOA. Because we use pulse profiles rather than individual pulses, the TOA is defined usually as the arrival time of the nearest pulse to the mid-point of the observation. As the pulses have a certain width, the TOA refers to some *fiducial point* on the profile. Ideally, this point coincides with the plane defined by the rotation and magnetic axes of the pulsar and the line of sight to the observer which is defined geometrically and independent of observing frequency or propagation effects.

The aim of pulsar timing is to count the number of neutron star rotations between two observations. Each TOA can therefore be assigned with a pulse number N which depends on rotation frequency ν and TOA t as

$$N = N_0 + \nu(t_0)(t - t_0) + \frac{1}{2}\dot{\nu}(t - t_0)^2 + \frac{1}{6}\ddot{\nu}(t - t_0)^3 + \cdots \quad (6)$$

where N_0 is the pulse number at the reference epoch t_0. If t_0 coincides with the arrival of a pulse and the pulsar spin-down (i.e. ν and $\dot{\nu}$) is known accurately, the pulses should appear at integer values of N when observed in an inertial reference frame. However, our observing frame is not inertial: we are using telescopes that are located on a rotating Earth orbiting the Sun. Before analysing TOAs measured with the observatory clock (topocentric arrival times), we need to transfer them to the centre of mass of the Solar System (solar system barycentre, SSB). To a very good approximation, the SSB is an inertial reference frame.

The time transformation also corrects for any relativistic time delay that occurs due to the presence of masses in the Solar System. An additional advantage of analysing these barycentric arrival times is that they can easily be combined with other TOAs measured at different observatories at different times.

Given a minimal set of starting parameters, a least squares fit is needed to match the measured arrival times to pulse numbers according to Equation (6). We minimise the expression

$$\chi^2 = \sum_i \left(\frac{N(t_i) - n_i}{\sigma_i} \right)^2 \quad (7)$$

where n_i is the nearest integer to $N(t_i)$ and σ_i is the TOA uncertainty in units of pulse period (turns).

The aim is to obtain a phase-coherent solution that accounts for every single rotation of the pulsar between two observations. One starts off with a small set of TOAs that were obtained sufficiently close in time so that the accumulated uncertainties in the starting parameters do not exceed one pulse period. Gradually, the data set is expanded, maintaining coherence in phase. When successful, post-fit residuals expressed in pulse phase show a Gaussian distribution around zero with a root mean square that is comparable to the TOA uncertainties (see Fig. 3). A good test for the quality of the TOAs and their fit is provided by creating a new set of mean residuals, each formed by averaging $n_{\rm avg}$ consecutive post-fit residuals. The root mean square calculated from the new set should decrease with $\sqrt{n_{\rm avg}}$ if no systematics are present.

After starting with fits for only period and pulse reference phase over some hours and days, longer time spans slowly require fits for parameters like spin frequency derivative(s) and position. Incorrect or incomplete timing models cause

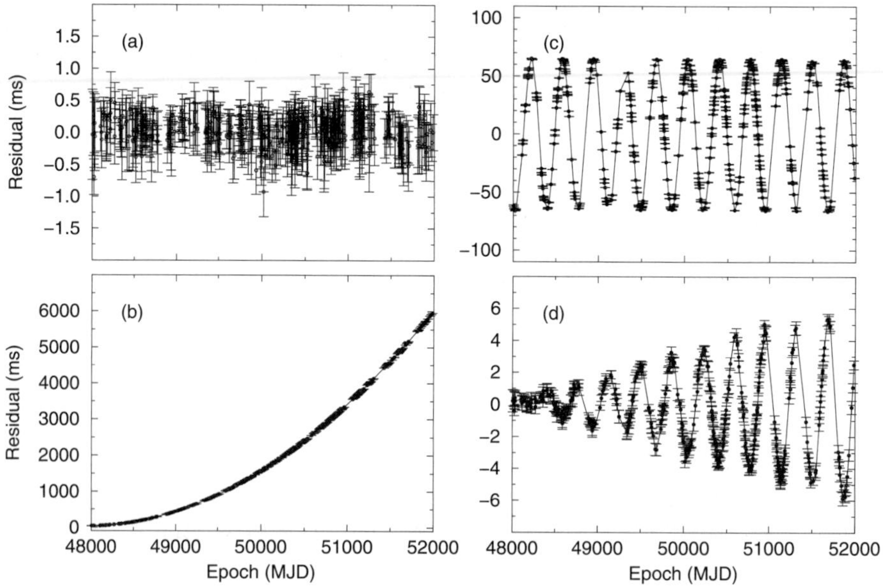

Figure 3: (a) Timing residuals for the 1.19 s pulsar B1133+16. A fit of a perfect timing model should result in randomly distributed residuals. (b) A parabolic increase in the residuals is obtained if $\dot{P}$ is underestimated, here by 4 per cent. (c) An offset in position (in this case a declination error of 1 arcmin) produces sinusoidal residuals with a period of 1 yr. (d) The effect of neglecting the pulsar's proper motion, in this case $\mu_{\rm T} = 380$ mas yr^{-1}. In all plots we have set the reference epoch for period and position to the first TOA at MJD 48000 to show the development of the amplitude of the various effects. Note the different scales on each of the vertical axes.

systematic structures in the post-fit residuals identifying the parameter that needs to be included or adjusted (see Figure 3). The precision of the parameters improves with length of the data span and the frequency of observation, but also with orbital coverage in the case of binary pulsars.

7. Binary Pulsars

Observations of pulsars in binary orbits show a periodic variation in pulse arrival time. The timing model therefore needs to incorporate the additional motion of the pulsar as it orbits the common centre of mass of the binary system. For non-relativistic binary systems, the orbit can be described using Kepler's laws. For a number of binary systems however, the Keplerian description of the orbit is not sufficient and relativistic corrections need to be applied.

Kepler's laws can be used to describe a binary system in terms of the five Keplerian parameters, shown schematically in Figure 4. These five parameters are required to refer the TOAs to the binary barycentre: (a) orbital period, $P_{\rm b}$; (b) projected semi-major orbital axis, $a_{\rm p} \sin i$ (see below); (c) orbital eccentricity, e; (d) longitude of periastron, ω; (e) the epoch of periastron passage, T_0.

For pulsars in close binary systems about white dwarfs, other neutron stars, or perhaps eventually black holes, relativistic effects due to strong gravitational fields and high orbital velocities produce observable signatures in the timing residuals. Even though GR appears to be the best description of the strong-field regime to date [17], alternative theories of gravity nevertheless should be considered and tested against it. A straightforward means of comparison is to parameterise the timing model in terms of the so-called 'post-Keplerian' (PK) parameters. For point masses with negligible spin contributions, the PK parameters in each theory should only be functions of the a priori unknown pulsar and companion mass, $M_{\rm p}$ and $M_{\rm c}$, and the easily measurable Keplerian parameters [18]. With the two masses as the only free parameters, an observation of two PK parameters will already determine the masses uniquely in the framework of the given theory. The measurement of a third or more PK parameters then provides a consistency check for the assumed theory.

The PK parameters are measured as additional parameters in a theory independent timing model which describes the pulse arrival times in a phenomenological way, using the Keplerian and PK parameters. The best timing model for describing relativistic binary pulsars is the Damour-Deruelle (DD) timing model [19, 20]. Applying this model to TOA measurements, the PK parameters are determined. They take different forms in different theories of gravity. In general relativity, the five most important PK parameters are given by (e.g., [18]):

$$\dot{\omega} = 3T_\odot^{2/3} \left(\frac{P_{\rm b}}{2\pi}\right)^{-5/3} \frac{1}{1-e^2} (M_{\rm p}+M_{\rm c})^{2/3}, \tag{8}$$

$$\gamma = T_\odot^{2/3} \left(\frac{P_{\rm b}}{2\pi}\right)^{1/3} e \frac{M_{\rm c}(M_{\rm p}+2M_{\rm c})}{(M_{\rm p}+M_{\rm c})^{4/3}}, \tag{9}$$

$$\dot{P}_{\rm b} = -\frac{192\pi}{5} T_\odot^{5/3} \left(\frac{P_{\rm b}}{2\pi}\right)^{-5/3} \frac{\left(1+\frac{73}{24}e^2+\frac{37}{96}e^4\right)}{(1-e^2)^{7/2}} \frac{M_{\rm p}M_{\rm c}}{(M_{\rm p}+M_{\rm c})^{1/3}}, \tag{10}$$

$$r = T_\odot M_{\rm c}, \tag{11}$$

$$s = T_\odot^{-1/3} \left(\frac{P_{\rm b}}{2\pi}\right)^{-2/3} x \frac{(M_{\rm p}+M_{\rm c})^{2/3}}{M_{\rm c}}, \tag{12}$$

where the masses M_p and M_c of pulsar and companion, respectively, are expressed in solar masses ($M_\odot$). We define the constant $T_\odot = GM_\odot/c^3 = 4.925490947\mu$s where G denotes the Newtonian constant of gravity and c the speed of light. The first PK parameter, $\dot{\omega}$, is the easiest to measure and describes the relativistic advance of periastron. It provides an immediate measurement of the total mass of the system, $(M_{\rm p}+M_{\rm c})$. The parameter γ denotes the amplitude of delays in

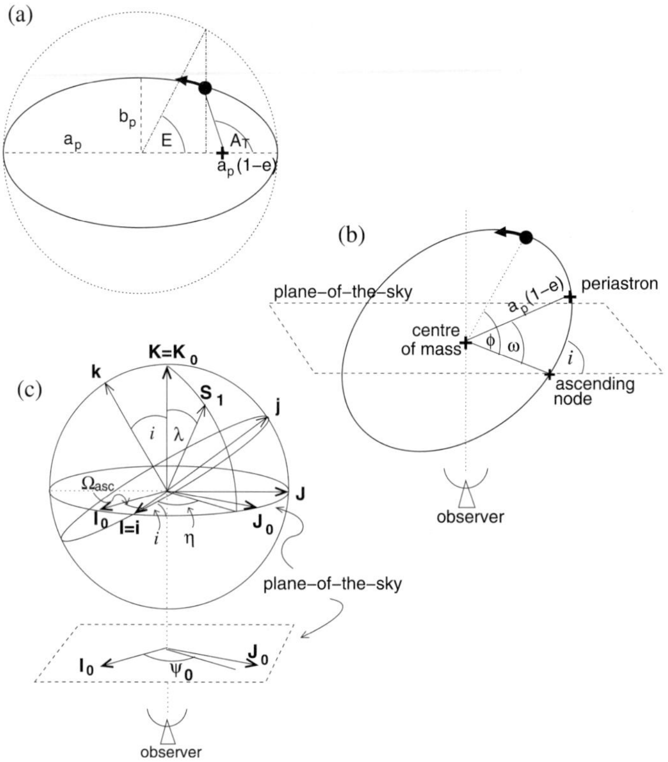

Figure 4: Definition of the orbital elements in a Keplerian orbit and the angles relating both the orbit and the pulsar to the observer's coordinate system and line of sight. (a) is drawn in the plane of the orbit; (b) shows the orbit inclined to the plane of the sky. The closest approach of the pulsar to the centre of mass of the binary system marks periastron, given by the longitude ω and a chosen epoch T_0 of its passage. The distance between centre of mass and periastron is given by $a_{\rm p}(1-e)$ where $a_{\rm p}$ is the semi-major axis of the orbital ellipse and e its eccentricity. (b) Usually, only the projection on the plane of the sky, $a_{\rm p} \sin i$, is measurable, where i is the orbital inclination defined as the angle between the orbital plane and the plane of the sky. The *true anomaly*, $A_{\rm T}$, and *eccentric anomaly*, E, are related to the *mean anomaly* by Kepler's law. The orbital phase of the pulsar Φ is measured relative to the ascending node. (c) The spatial orientation of the pulsar's spin-vector, $\mathbf{S_1}$, is given by the angles λ and η in the coordinate system shown as defined by Damour and Taylor (1992). The angle $\Omega_{\rm asc}$ gives the longitude of ascending node in the plane of the sky.

arrival times caused by the varying effects of the gravitational redshift and time dilation (second order Doppler) as the pulsar moves in its elliptical orbit at varying distances from the companion and with varying speeds. The decay of the orbit due to gravitational wave damping is expressed by the change in orbital period, $\dot{P}_b$. The other two parameters, r and s, are related to the Shapiro delay caused by the gravitational field of the companion. These parameters are only measurable, depending on timing precision, if the orbit is seen nearly edge-on.

The PK parameters listed above are those which have been measured in binary systems to date. However, the list can be extended (see [18]) if the binary system is extreme enough. We expect the double pulsar to be such a system where never before measured PK parameters will be needed to describe the observations adequately (see Section 10).

8. The double pulsar – a magnificent laboratory

Our team discovered the 22.8-ms pulsar J0737−3039 in April 2003 [21] in an extension to the hugely successful Parkes Multi-beam survey [22]. It was soon found to be a member of the most extreme relativistic binary system ever discovered: its short orbital period ($P_b = 2.4$ hrs) is combined with a remarkably high value of periastron advance ($\dot{\omega} = 16.9\,\mathrm{deg\ yr}^{-1}$, i.e. four times larger than for the Hulse-Taylor pulsar PSR B1913+16). This large precession of the orbit was measurable after only a few days of observations. The system parameters predict that the two members of the binary system will coalesce on a short time scale of only ~ 85 Myr. This boosts the hopes for detecting a merger of two neutron stars with first-generation ground-based gravitational wave detectors by a factor of 5 to 10 compared to previous estimates based on only the double neutron stars B1534+12 and B1913+16 [21, 23].

In October 2003, we detected radio pulses from the second neutron star [24]. The reason why signals from the 2.8-s pulsar companion (now called PSR J0737−3039B, hereafter "B") to the millisecond pulsar (now called PSR J0737−3039A, hereafter "A") had not been found earlier, became clear when it was realized that B was only bright for two short parts of the orbit. For the remainder of the orbit, the pulsar B is extremely weak and only detectable with the most sensitive equipment. The detection of a young companion B around an old millisecond pulsar A and their position in the $P - \dot{P}$-diagram (see Figure 2) confirms the evolution scenario proposed for recycled pulsars (see Section 5) and provides a truly unique testbed for relativistic gravity and also plasma physics.

8.1. A laboratory for plasma physics

The double pulsar is not only a superb test-bed for relativistic gravity, but it also provides an unprecedented opportunity to probe the workings of pulsars. The pulse emission from B is strongly modulated with orbital phase, most probably as a consequence of the penetration of the A's wind into B's magnetosphere. Figure 5 shows the pulse intensity for B as a function of pulse phase and orbital longitude for three radio frequencies. The first burst of strong emission, centred near orbital longitude 210 deg, covers about 13 min of the orbit, while the second burst, centred near longitude 280 deg, is shorter and lasts only for about 8 min. This pattern is stable over successive orbits and obviously frequency independent over the range probed. Deep integrations reveal other orbital phases, where B is visible but much weaker than during the two main burst periods. The figure also shows that not only does the pulse intensity change with orbital phase, but that the pulse shape changes as well. At the start of the first burst the pulse has a strong trailing component and a weaker leading component which dies out in the later phases of the burst. In the second burst, there are two components of more equal amplitude. This is the first time that profile changes are observed that clearly depend on orbital phase. Decoding this pattern as the orbit precesses due to relativistic effects and the system is viewed from different directions, offers a unique chance to probe the magnetosphere. Indeed, as discussed later, the "light-curve" of B is changing with time, probably due to the effects of geodetic precession.

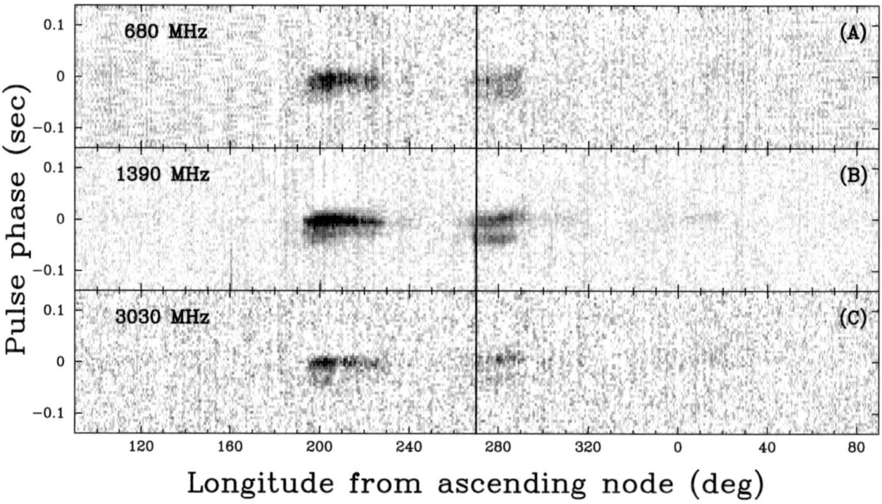

Figure 5: Grey-scale images showing the pulse of PSR J0737–3039B as a function of orbital phase at three observing frequencies (Lyne *et al.* 2004).

It is important to note that by simply seeing B functioning as a radio pulsar, albeit with orbital phases of rather weak emission, confirms our ideas about the location of the origin of radio emission: The fact that B is still emitting, despite the loss of most of its magnetosphere due to A's wind, indicates that the fundamental processes producing radio emission are likely to occur close to the neutron star surface – in accordance with emission heights determined for normal radio pulsars (see Section 3).

The quenching or attenuation of B's radio emission for most of its orbit is only part of the interaction between A and B that is observed. For about 27 seconds of the orbit, A's emission is eclipsed when A is lined up behind B at superior conjunction (Fig. 6). At that moment, the pulses of A pass in about 30,000 km distance to the surface of B. It appears that the magnetospheric transmission for A's emission is modulated during the rotation of B, depending on the relative orientation of the spin-axis of B to A and our line-of-sight. Indeed, a modulation of the light-curve inside the eclipse region consistent with B's (full and half) rotation period is observed [25] (Fig. 7).

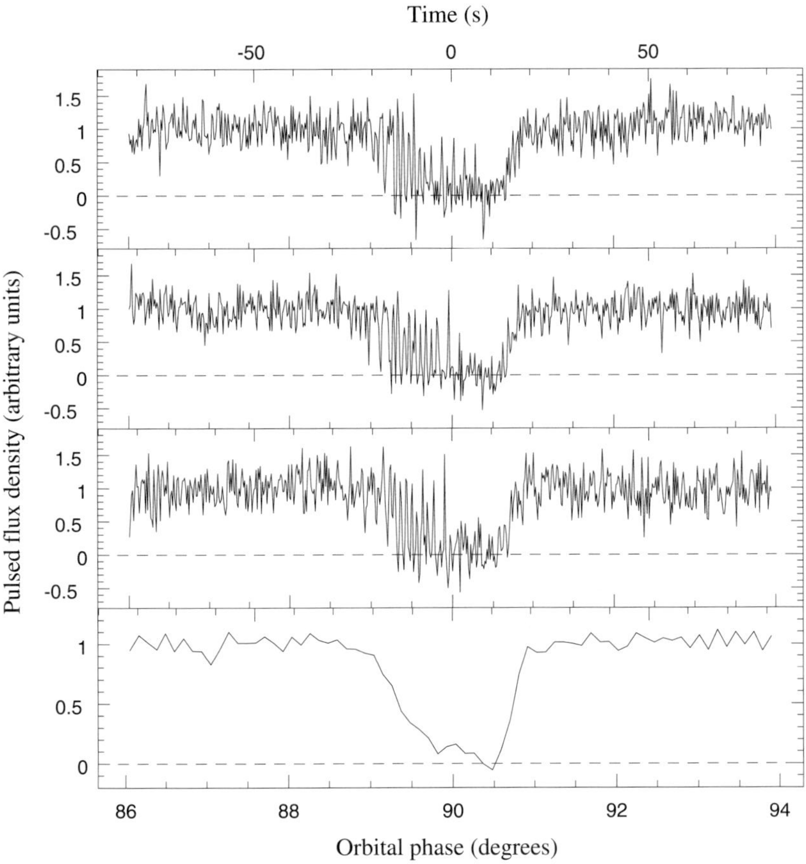

Figure 6: The pulsed flux density of A versus time (with respect to superior conjunction) and orbital phase for (top three panels) the three eclipses in the 820-MHz observation and (bottom panel) all three eclipses summed (McLaughlin et al. 2004b). In the individual eclipse light curves, every 12 pulses have been averaged for an effective time resolution of ~ 0.27 s. Every 100 pulses have been averaged to create the lower, composite light curve for an effective time resolution of ~ 2.3 s. Pulsed flux densities have been normalized such that the pre-eclipse average flux density is unity.

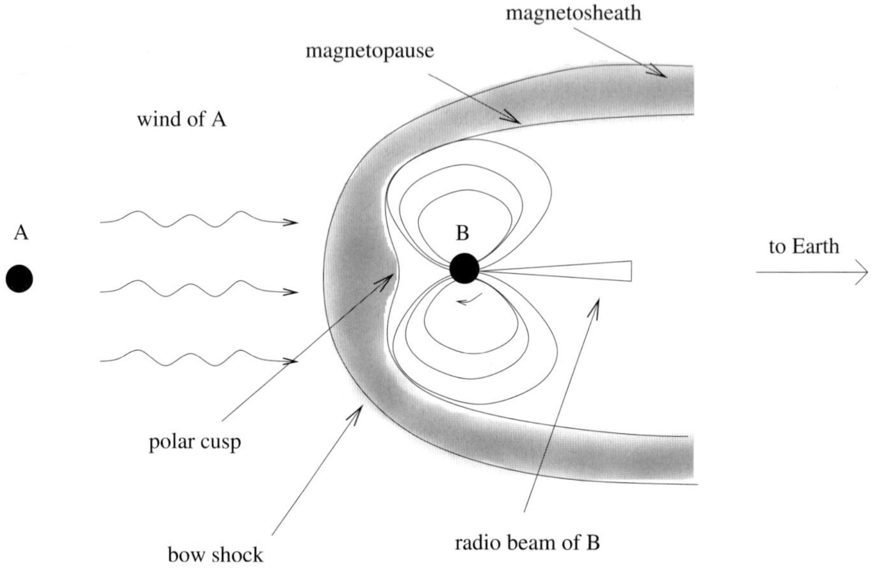

Figure 7: Cartoon (not to scale) showing the interaction between the relativistic wind of A and the magnetosphere of B when the radio beam of B is pointing towards the Earth (from McLaughlin et al. 2004b).

Perhaps even more exciting is the discovered evidence that A's radiation has some direct impact on the radiation pattern of B. Figure 8 shows a blow-up of B's emission at orbital phases where B is strongest. At the right orientation angles, a drifting sub-pulse pattern emerges that coincides with the arrival times of A's pulses at B [26]. This is the first time pulsar emission is observed to be triggered by some external force, and it is likely that this will help us to understand the conditions and on-set of pulsar emission in general.

8.2. A laboratory for strong-field gravity

Since neutron stars are very compact massive objects, the double pulsar (and other double neutron star systems) can be considered as almost ideal point sources for testing theories of gravity in the strong-gravitational-field limit. Timing observations of PSR J0737−3039A/B have been undertaken using the 64-m Parkes radio telescope in New South Wales, the 76-m Lovell radio telescope at Jodrell Bank Observatory, UK, and the 100-m Green Bank Telescope in West Virginia, between 2003 April and 2006 January. A total of 131,416 pulse times-of-arrival (TOAs) were measured for A while 507 TOAs were obtained for B using synthetic noise-free profile templates which are different for different frequency bands and, for B, functions of orbital phase and time. Pulsar and binary system parameters derived

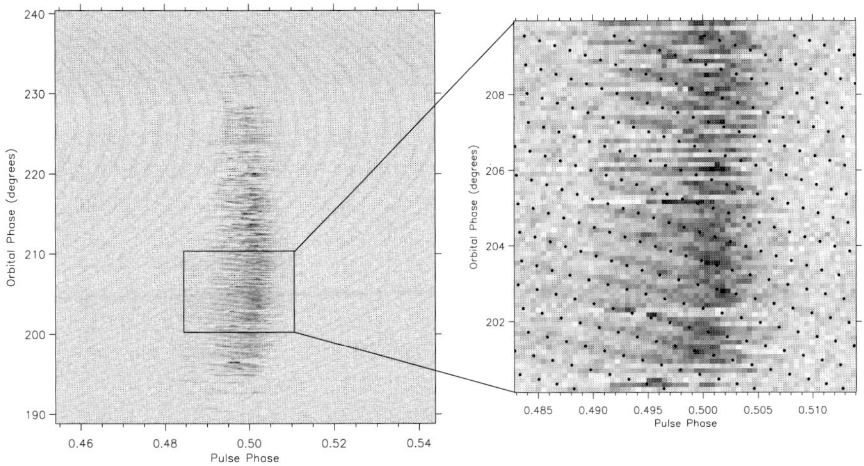

Figure 8: Observations of single pulses of B at 820 MHz for orbital phases 190–240 deg (only 10% of the pulse period is shown). Drifting features are present through most of these data, but are particularly obvious from orbital phases ~ 200–210 deg which is enlarged on the right. Single pulses of A can be seen in the background of the left figure, where differential Doppler shifts from the orbital motion result in different apparent pulse periods and hence drifting patterns. The expanded view on the right is overlayed with dots marking the arrival of pulses of A at the centre of B, coinciding with the observed drift pattern in B (McLaughlin et al. 2004a).

from these TOAs are listed in Table 1 at the end of this paper. Because of its narrower and more stable pulse profile, TOAs from A have a much higher precision than those from B and hence are used to determine the position, proper motion and main orbital parameters of the system. For B, the only fitted parameters were the pulse phase, the pulsar spin frequency, $\nu \equiv 1/P$, its first time-derivative $\dot{\nu}$ and the projected semi-major axis, $x_{\rm B} \equiv (a_{\rm B}/c)\sin i$.

Previous observations of PSR J0737−3039A/B [21, 24] resulted in the measurement of R and four PK parameters: the rate of periastron advance $\dot{\omega}$, the gravitational redshift and time dilation parameter γ, and the Shapiro-delay parameters r and s. For the latest timing results [17] the measurement precision for these parameters has increased dramatically. Also, we have now measured the orbital decay, $\dot{P}_{\rm b}$, giving a total of five PK parameters for the system. Together with the mass ratio R, the measurements provide four independent tests of GR, more than for any other known system. The value of $\dot{P}_{\rm b}$, measured at the 1.4% level after only 2.5 years of timing, corresponds to a shrinkage of the orbit at a rate of 7 mm per day. One can display these tests elegantly in a "mass-mass" diagram as shown in Figure 9. Measurement of the PK parameters gives curves on this diagram that are in general different for different theories of gravity but which should intersect in a single point, i.e., at a pair of mass values, if the theory is valid [18].

In addition to tests enabled by the PK parameters, the access to the orbit of both neutron stars – by timing A *and* B – provides yet another constraint on gravitational theories that is qualitatively different from what has been possible with previously known double neutron stars: using Kepler's third law, the measurement of the projected semi-major axes of both orbits yields the mass ratio,

$$R(M_A, M_B) \equiv M_A/M_B = x_B/x_A. \tag{13}$$

For every realistic theory of gravity, we can expect R to follow this simple relation [18], at least to 1PN order. Most importantly, the R value is not only theory-independent, but also independent of strong-field (self-field) effects which is not the case for the PK parameters. In other words, any combination of masses derived from the PK parameters *must* be consistent with the mass ratio. The ability to measure this quantity provides therefore an important and unique constraint. With five PK parameters already available, this additional constraint also makes the double pulsar the most overdetermined system to date where the most relativistic effects can be studied in the strong-field limit.

Figure 9 shows that all measured constraints are consistent with GR. The most precisely measured PK parameter currently available is the precession of the longitude of periastron, $\dot{\omega}$. We can combine this with the theory-independent mass ratio R to derive the masses given by the intersection region of their curves: $m_{\rm A} = 1.3381 \pm 0.0007$ M$_\odot$ and $m_{\rm B} = 1.2489 \pm 0.0007$ M$_\odot$. Assuming GR and using these masses and the Keplerian parameters, we can predict values for the remaining PK parameters. Table 2 at the end of this paper lists results for the four independent tests that are currently available. The Shapiro delay (Figure 10) gives the most precise test, with $s_{\rm obs}/s_{\rm pred} = 0.99987 \pm 0.00050$. This is by far the best test of GR in the strong-field limit, having a higher precision than the test based on the observed orbit decay in the PSR B1913+16 system with a 30-year data span [30]. As for the PSR B1534+12 system [31], the PSR J0737−3039A/B Shapiro-delay test is complementary to that of B1913+16 since it is not based on predictions relating to emission of gravitational radiation from the system [32]. Most importantly, the four tests of GR presented here are qualitatively different from all previous tests because they include one constraint (R) that is independent of the assumed theory of gravity at the 1PN order. As a result, for any theory of gravity, the intersection point is expected to lie on the mass ratio line in Figure 9. GR also passes this additional constraint with the best precision so far.

8.2.1. A modification of the DD timing model. In order to study such possible effects and the performance of the standard timing software TEMPO[2] and its implementation of the DD timing model, we have made detailed simulations. Producing fake TOAs for a J0737−3039-like system, we varied the input parameter

[2] `http://pulsar.princeton.edu/tempo/`

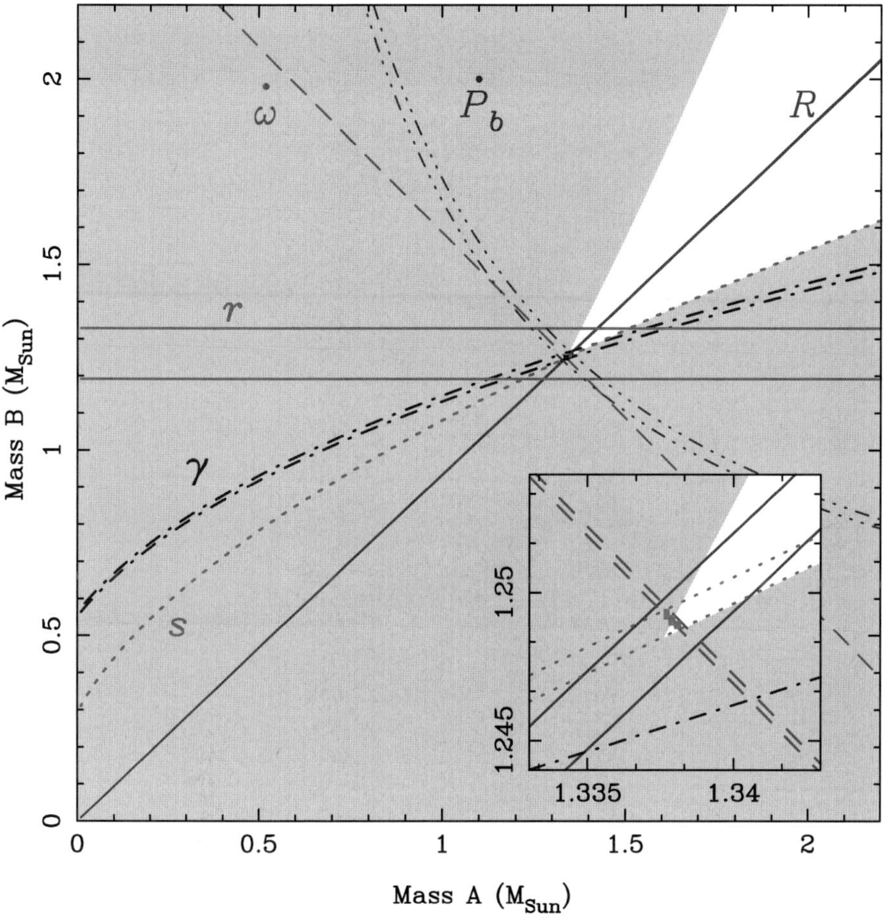

Figure 9: 'Mass–mass' diagram showing the observational constraints on the masses of the neutron stars in the double pulsar system J0737–3039 (Kramer et al. 2006). The shaded regions are those that are excluded by the Keplerian mass functions of the two pulsars. Further constraints are shown as pairs of lines enclosing permitted regions as given by the observed mass ratio and PK parameters as predicted by general relativity. Inset is an enlarged view of the small square encompassing the intersection of these constraints (see text).

as $0.9 \leq s \leq 1.0$ and the assumed timing precision. For small TOA uncertainties, we can always recover the original s value by fitting the DD-model using TEMPO. However, comparing the standard TEMPO error estimates for s and r with estimates obtained from studying a corresponding χ^2 plane, the symmetric

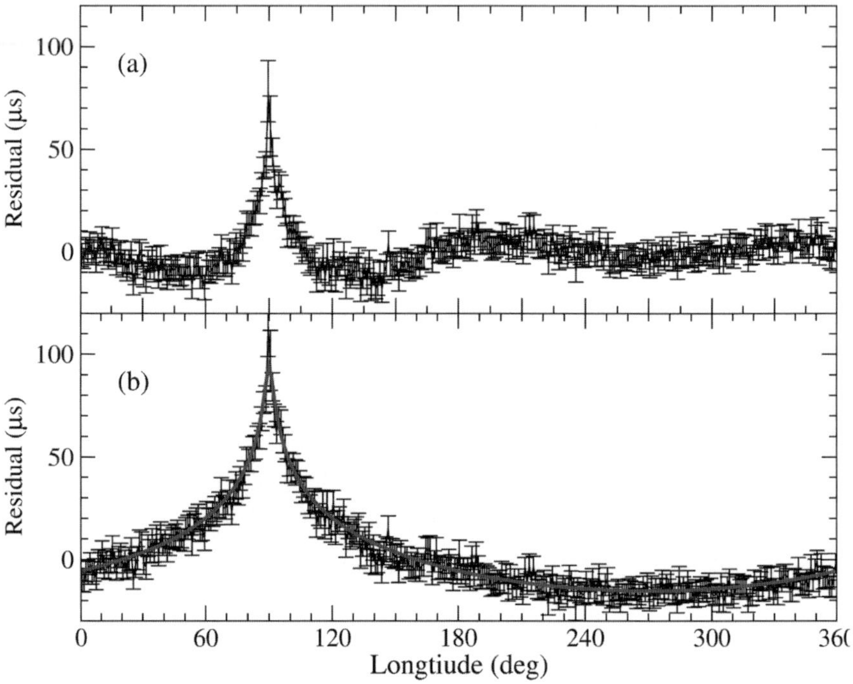

Figure 10: The effect of the Shapiro delay caused by the gravitational potential of B seen in the timing residuals of A. (a) Observed timing residuals after a fit of all model parameters given in Table 1 *except* the Shapiro-delay terms r and s which were set to zero. (b) Residuals illustrating the full Shapiro delay, obtained by holding all parameters to their values given in Table 1, except the Shapiro delay terms which were set to zero. The line shows the predicted delay at the centre of the data span. In both cases, residuals were averaged in 1° bins of longitude (Kramer et al. 2006).

error bars given by TEMPO do not always correspond to the true uncertainties reflected by non-symmetric χ^2-contours if the TOA uncertainty is too large. This potential problem due to the non-linearity of the fitted parameters and correlations of the Shapiro delay parameters with the Römer delay in the orbit is well known. Hence, one usually explores an χ^2-plane evenly sampled in $\sqrt{1-s^2}$ and r to obtain reliable values and error estimates (see, e.g., [34]). Increasing the TOA uncertainties, numerical fits to the fake TOAs assuming a $\sin i$ very close to unity (e.g., $\sin i = 0.99999$ or $i = 89.2$) sometimes results in fits with $s > 1$ due to numerical uncertainties. In order to remedy this situation we developed a modification of the DD timing model following a suggestion by Thibault Damour (priv. communication).

In the DD model, we fit for r and s which in GR becomes $s = \sin i$ where r is identical with the companion mass apart from a constant factor, $T_\odot$ (see Eqns. 11 and 12). In the new model, called DDS (for DD-Shapiro, see [35]), we write

$$s = 1 - e^{-z_s} \tag{14}$$

where z_s replaces s as our new fit parameter. It follows that

$$z_s = -\ln(1 - s). \tag{15}$$

The advantage becomes apparent when we compare this expression to the Shapiro delay term, Δ_s in the timing formula, in particular when comparing it to low-eccentricity pulsars for which (e.g., [36])

$$\Delta_s = -2r \ln(1 - s \sin \Phi) \tag{16}$$

where Φ is the orbital phase measured from the ascending node. At $\Phi = \pi/2$, the maximum delay is obtained

$$\Delta_s^{max} = -2r \ln(1 - s) \tag{17}$$

which has obvious similarities to our definition of z_s,

$$\Delta_s^{max} = 2 \, r \, z_s \tag{18}$$

or

$$z_s = \Delta_s^{max}/2r. \tag{19}$$

Due to the nature of this simple transformation, a fit of the TOAs to the DD and DDS models always produces the same results. In addition, however, at large inclination angles the uncertainties on z_s derived by TEMPO are still consistent with those obtained from studies of corresponding χ^2 hyperspheres, often removing the need for the often computationally expensive calculation of the χ^2 plane. Using the DDS model, it is in particular impossible that numerical uncertainties lead to fit results which in GR correspond to values $\sin i > 1.0$. We are aware that the DDS model therefore represents a restriction of the parameter space which may be allowed by alternative theories of gravity.

The application of the DD and DDS model to the real TOAs produces consistent results and verifies the previous findings that s is significantly lower than the scintillation results which are consistent with our measurement only at the 3-σ level. Scintillation observations over the whole orbit have also been used to deduce the system transverse velocity. Ransom et al. [37] derive a value of 141 ± 8.5 km s^{-1} while Coles et al. [33] obtain 66 ± 15 km s^{-1} after considering the effect of anisotropy in the scattering screen. Both of these values are in stark contrast to the value of 10 ± 1 km s^{-1} (relative to the solar system barycentre) obtained from pulsar timing (Table 1). We note that the scintillation-based velocity depends on a number of assumptions about the properties of the effective scattering screen. In contrast, the proper motion measurement has a clear and unambiguous timing signature, although the transverse velocity itself scales with the pulsar distance.

Even allowing that unmodeled effects of Earth motion could affect the published scintillation velocities by about 30 km s^{-1}, the dispersion-based distance would need to be underestimated by a factor of several to make the velocities consistent. We believe this is very unlikely, particularly as the tentative detection of a parallax gives us some confidence in the dispersion-based distance estimate. Hence, our timing results for both inclination angle and transverse velocity are less susceptible to systematic errors and are therefore more secure than those based on scintillation.

We have studied other possible effects affecting the timing results such as a possible variation of the dispersion measure as a function of orbital phase. However, the non-detection of any such effect leads us to the conclusion that in contrast unmodeled effects may have altered the scintillation results and the derived uncertainties on the inclination angle. An exciting explanation for the discrepancy may be that the emission of A suffers measurable refraction while propagating through the magnetosphere of B. If that were indeed the case, we would have a direct handle onto the magneto-ionic properties of B's magnetosphere for the first time, e.g., corresponding plasma densities in B's magnetosphere would need to be relatively large.

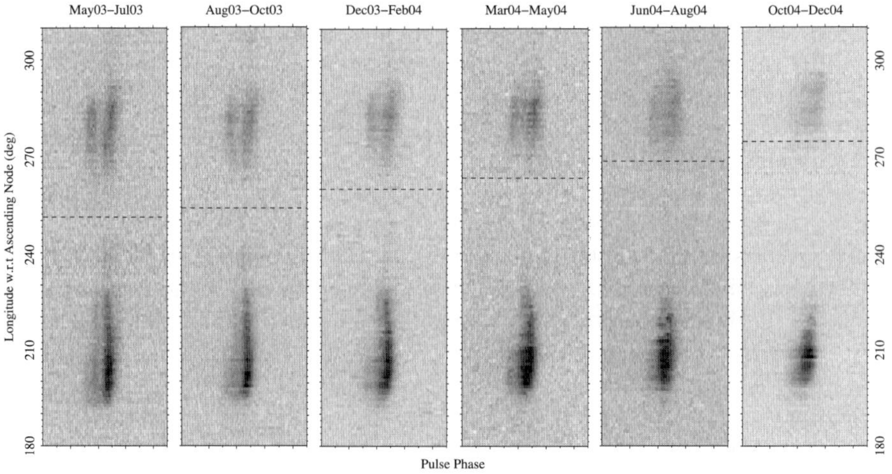

Figure 11: The emission of B at 1390 MHz as a function of the orbital longitude (vertical axis) and for the pulse phase range 0.18–0.27 (horizontal axis) as in Figure 5. Each panel was obtained by adding all the data in the $\sim$ 3 months period indicated at the top of each panel. The dashed lines represent the position of the periastron at the given epoch. A change in the visibility pattern of B is clearly visible and most likely caused by geodetic precession (Burgay et al. 2005).

The short eclipses in A's emission already indicate that we are observing the system almost completely edge-on. This is confirmed independently by measuring a Shapiro delay and inferring (in GR) that the inclination angle i is indeed close to 90°. Other methods using the variation of the pulsars's intensity due to the turbulent interstellar medium are available for an independent measurement of the orbital orientation. Based on such scintillation observations of both pulsars over the short time interval when A is close to superior conjunction, Coles et al. [33] derived a value, the orbital inclination angle i, that is very close to 90°, i.e. they derive $|i - 90°| = 0°.29 \pm 0°.14$. In GR, the parameter s can be identified with $\sin i$ where i is the inclination angle of the orbit. The value of s derived from our timing observations (Table 1) corresponds to $i = 88°.69^{+0°.50}_{-0°.76}$. Comparing the two methods, one notes that the scintillation results are based on correlating the scintillation properties of A and B over the short time-span of the orbital motion when they are in conjunction to the observer. In contrast, the measurement of the inclination angle from timing measurements results from detecting significant harmonic structure in the post-fit residuals after parts of the Shapiro delay are absorbed in the fit for the Römer delay, i.e. the light travel time across the orbit. As shown in Figure 10, these structures are present throughout the whole orbit, so that the results from timing measurements may be expected to be more reliable. However, as all TOAs are associated with uncertainties, we need to make sure that a multi-parameter least-square fit of the DD model will reproduce the correct value of the PK parameters s and r despite possible numerical effects.

8.2.2. Effects of geodetic precession. The measurement of the times-of-arrival (TOAs) are obtained with a standard "template matching" procedure that involves a cross-correlation of the observed pulse profile with high signal-to-noise ratio template (e.g., [38]). Any change in the pulse profile could lead to systematic variations in the measured TOAs. We performed detailed studies of the profiles of A and B to investigate any possible profile changes with time as such as expected from another effect predicted by GR.

In GR, the proper reference frame of a freely falling object suffers a precession with respect to a distant observer, called geodetic precession. In a binary pulsar system this geodetic precession leads to a relativistic spin-orbit coupling, analogous to spin-orbit coupling in atomic physics [39]. As a consequence, both pulsar spins precess about the total angular momentum, changing the relative orientation of the pulsars to one another and toward Earth. Since the orbital angular momentum is much larger than the pulsars' angular momenta, the total angular momentum is effectively represented by the orbital angular momentum. The precession rate [40] depends on the period and the eccentricity of the orbit as well as the masses of A and B. With the orbital parameters of the double pulsar, GR predicts precession periods of only 75 yr for A and 71 yr for B.

Geodetic precession has a direct effect on the timing as it causes the polar angles of the spins and hence the effects of aberration to change with time [18]. These changes modify the *observed* orbital parameters, like projected semi-major axis and eccentricity, which differ from the *intrinsic* values by an aberration dependent term, potentially allowing us to infer the system geometry (see Section 10).

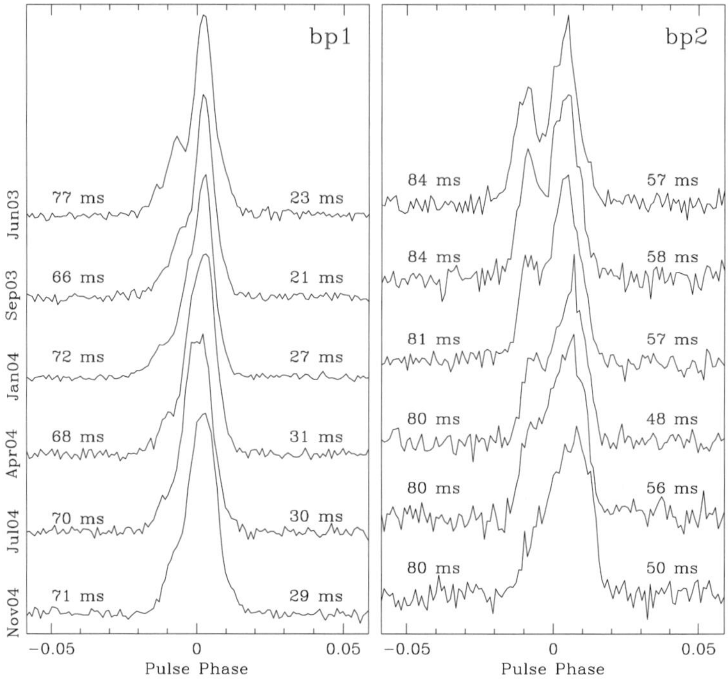

Figure 12: Pulse shapes of B in the two bright phases (see previous figure) as a function of time, again in three month intervals. Again, a systematic change in the pulse profile is clearly visible (Burgay et al. 2005).

Extracting the signature of these effects in the timing data is a goal for the years to come. Other consequences of geodetic precession can be expected to be detected much sooner and are directly relevant for the timing of A and B. These arise from variations in the pulse shape due to changing cuts through the emission beam as the pulsar spin axes precess. Moreover, geodetic precession also leads to a change in the relative alignment of the pulsar magnetospheres, so that the visibility pattern and even the profile of B should vary due to these changes as well.

Indeed, studies of the profile evolution of B [41] reveal a clear evolution of B's emission on orbital and secular time-scales. The light-curves of B (i.e. the visibility of B versus orbital phase) undergo clear changes (Figure 11) while the profile of B as observed in the bright orbital phases is clearly changing also with time (Figure 12). These phenomena are probably caused by a changing magnetospheric interaction due to geometry variations resulting from geodetic precession. In any case, these changes require sophisticated timing analysis techniques.

For the timing of B, because of the orbital and secular dependence of its pulse profile, different templates were used for different orbital phases and different

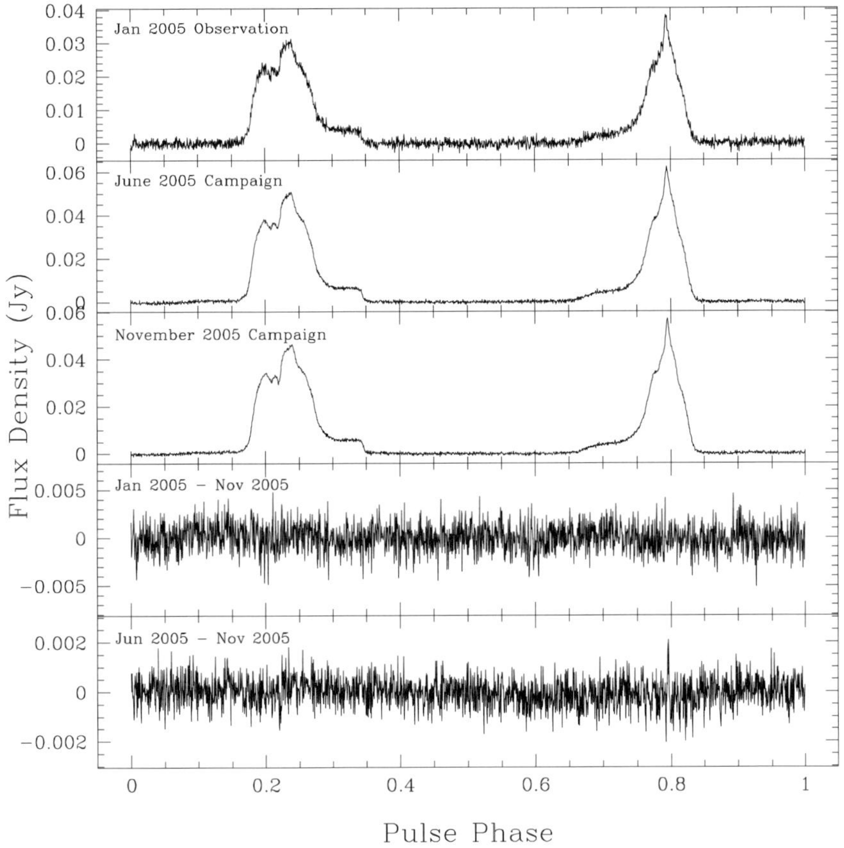

Figure 13: Pulse profiles of A observed at three different epochs. Subtracting one profile from the others does not reveal any significant changes as demonstrated in the two bottom plots. Figure provided by Rob Ferdman & Ingrid Stairs.

epochs. A matrix of B templates was constructed, dividing the data set into 3-month intervals in epoch and 5-minute intervals in orbital phase. The results for the 29 orbital phase bins were studied, and it was noticed that, while the profile changes dramatically and quickly during the two prominent bright phases, the profile shape is simpler and more stable at orbital phases when the pulsar is weak. In the final timing analysis for pulsar B, we therefore omitted data from the two very bright orbital phases. We also used an unweighted fit to avoid biasing the fit toward remaining brighter orbital phases.

Since the overall precision of our tests of GR is currently limited by our ability to measure $x_{\rm B}$ and hence the mass ratio $R \equiv m_{\rm A}/m_{\rm B} = x_{\rm B}/x_{\rm A}$, we adopted the following strategy to obtain the best possible accuracy for this parameter. We used the whole TOA data set for B in order to measure B's spin parameters P and $\dot{P}$, given in Table 1 of the main paper. These parameters were then kept fixed for a separate analysis of two concentrated 5-day observing sessions. On the timescale of the long-term profile evolution of B, each 5-day session represents a single-epoch experiment and hence requires only two sets of profile templates. The value of $x_{\rm B}$ obtained from a fit of this parameter only to the two 5-day sessions is presented in Table 1 of the main paper.

The study of the profile evolution of A [42, 17] did not lead to the detection of any profile change (see Figure 13). This present non-detection greatly simplifies the timing of A but does not exclude the possibility that changes may happen in the future. While the effects of geodetic precession could be small due to a near alignment of pulsar A's spin and the orbital momentum vector, the results could also be explained by observing the system at a particular precession phase. While this case appears to be relatively unlikely, it must not be excluded as such a situation had indeed occurred for PSR B1913+16 [43]. Indeed, a modelling of the results suggests that this present non-detection of profile changes is consistent with a rather wide range of possible system geometries. One conclusion that can be drawn is that the observations are inconsistent with the large profile changes that had been predicted by some models [44]. Fortunately, independent information is available that suggests that the alignment angle, and hence the impact of geodetic precession, may indeed be very small. This information is derived from a study of the evolution of the system based on the rather small transverse velocity of the double pulsar.

8.3. Space-motion and evolution of the double pulsar

The timing results indicate that the space velocity of the double pulsar system is surprisingly small. Based on the measured dispersion measure and a model for the Galactic electron distribution [29], PSR J0737−3039A/B is estimated to be about 500 pc from the Earth. From the timing data we have measured a marginally significant value for the annual parallax, 3 ± 2 mas, corresponding to a distance of $200 - 1000$ pc (Table 1), which is consistent with the dispersion-based distance that was also used for studies of detection rates in gravitational wave detectors [21]. The observed proper motion of the system (Table 1) then implies a transverse space velocity of only 10 km s^{-1} with a velocity vector parallel to the Galactic plane. With this information, Stairs et al. (2006) [45] examined the history and formation of this system, determining estimates of the pre-supernova companion mass, orbital separation, supernova kick and misalignment angle between the pre- and post-supernova orbital planes. One of the surprising results is that the progenitor to the recently formed B was probably less than $2\,M_\odot$, lending credence to suggestions that this object may not have formed in a normal core-collapse supernova. They conclude that it therefore must be possible, in at least some cases, for low-mass helium stars to undergo Supernova explosions, and that there must be a range of progenitor types that can produce double neutron star systems. The relative frequency of the different types must depend on the initial mass function and ranges of binary orbits. This will have implications for the number of double neutron systems in the Galaxy, the retention of neutron stars in globular clusters and for the apparent dearth of isolated mildly recycled pulsars ejected from unbinding second SN explosions.

The study of the double pulsar evolution [45] also suggests that the kick velocity was rather small and that the misalignment angle between the spin of pulsar A and the total angular momentum vector (after the second supernova explosion) is probably much less than 10°. In this case, the expected impact of geodetic precession on pulsar A's profile is rather small, consistent with the present observational evidence. This makes us confident that high-precision timing observations of A will continue for quite a while (in contrast to, for instance, the Hulse-Taylor pulsar which may disappear as a radio source at about 2025 [43]). At the same time, the small velocity of the double pulsar system is also extremely good news for tests of alternative theories of gravity where the measurement of the orbital decay is extremely useful.

9. Orbital decay measurement & Alternative theories of gravity

Because the measured uncertainty in $\dot{P}_{\rm b}$ decreases approximately as $T^{-2.5}$, where T is the data span, we expect to improve our test of the radiative aspect of the system to the 0.1% level or better in about five years' time. For the PSR B1913+16 and PSR B1534+12 systems, the precision of the GR test based on the orbit-decay rate is severely limited both by the uncertainty in the differential acceleration of the Sun and the binary system in the Galactic gravitational potential and the uncertainty in pulsar distance [46, 31]. For PSR J0737−3039A/B, both of these corrections are very much smaller than for these other systems. Based on the measured dispersion measure and a model for the Galactic electron distribution [29], PSR J0737−3039A/B is estimated to be about 500 pc from the Earth. From the timing data we have measured a marginally significant value for the annual parallax, 3 ± 2 mas, corresponding to a distance of $200 - 1000$ pc (Table 1), which is consistent with the dispersion-based distance that was also used for studies of detection rates in gravitational wave detectors [21]. The observed proper motion of the system (Table 1) and differential acceleration in the Galactic potential [47] then imply a kinematic correction to $\dot{P}_{\rm b}$ at the 0.02% level or less. Independent distance estimates also can be expected from measurements of the annual parallax by Very Long Baseline Interferometry (VLBI) observations, allowing a secure compensation for this already small effect. A measurement of $\dot{P}_{\rm b}$ at the 0.02% level or better will provide stringent tests for alternative theories of gravity as many are, for instance, predicting a significant amount of gravitational *dipole* radiation. Hence, a confirmation that the observations of the double pulsar are consistent with emission of gravitational *quadrupole* radiation to a very high level of precision promises to put limits on some scalar-tensor theories that will surpass even the best current Solar-system tests [48].

Timing parameter	PSR J0737−3039A	PSR J0737−3039B
Right Ascension α	$07^h37^m51^s.24927(3)$	–
Declination δ	$-30^\circ 39' 40''.7195(5)$	–
Proper motion in the RA direction (mas yr^{-1})	−3.3(4)	–
Proper motion in Declination (mas yr^{-1})	2.6(5)	–
Parallax, π (mas)	3(2)	–
Spin frequency ν (Hz)	44.054069392744(2)	0.36056035506(1)
Spin frequency derivative $\dot{\nu}$ (s^{-2})	$-3.4156(1) \times 10^{-15}$	$-0.116(1) \times 10^{-15}$
Timing Epoch (MJD)	53156.0	53156.0
Dispersion measure DM ($cm^{-3}pc$)	48.920(5)	–
Orbital period P_b (day)	0.10225156248(5)	–
Eccentricity e	0.0877775(9)	–
Projected semi-major axis $x = (a/c)\sin i$ (s)	1.415032(1)	1.5161(16)
Longitude of periastron ω (deg)	87.0331(8)	87.0331 + 180.0
Epoch of periastron T_0 (MJD)	53155.9074280(2)	–
Advance of periastron $\dot{\omega}$ (deg/yr)	16.89947(68)	[16.96(5)]
Gravitational redshift parameter γ (ms)	0.3856(26)	–
Shapiro delay parameter s	0.99974(−39, +16)	–
Shapiro delay parameter r (μs)	6.21(33)	–
Orbital period derivative $\dot{P}_b$	$-1.252(17) \times 10^{-12}$	–
Timing data span (MJD)	52760 – 53736	52760 – 53736
RMS timing residual σ (μsec)	54	2169
Total proper motion (mas yr^{-1})	4.2(4)	
Distance d(DM) (pc)	∼ 500	
Distance $d(\pi)$ (pc)	200 – 1000	
Transverse velocity (d = 500 pc) (km s^{-1})	10(1)	
Orbital inclination angle (deg)	88.69(-76,+50)	
Mass function ($M_\odot$)	0.29096571(87)	0.3579(11)
Mass ratio, R	1.0714(11)	
Total system mass ($M_\odot$)	2.58708(16)	
Neutron star mass ($m_\odot$)	1.3381(7)	1.2489(7)

Table 1: Parameters for PSR J0737−3039A (A) and PSR J0737−3039B (B) as measured by Kramer et al. (2006). The values were derived from pulse timing observations using the DD and DDS models of the timing analysis program TEMPO [27] and the Jet Propulsion Laboratory DE405 planetary ephemeris [28]. Estimated uncertainties, given in parentheses after the values, refer to the least significant digit of the tabulated value and are twice the formal 1-σ values given by TEMPO. The positional parameters are in the DE405 reference frame which is close to that of the International Celestial Reference System. Pulsar spin frequencies $\nu \equiv 1/P$ are in barycentric dynamical time (TDB) units at the timing epoch quoted in Modified Julian Days. The five Keplerian binary parameters (P_b, e, ω, T_0, and x) are derived for pulsar A. The first four of these (with an offset of 180° added to ω) and the position parameters were assumed when fitting for B's parameters. Five post-Keplerian parameters have now been measured. An independent fit of $\dot{\omega}$ for B yielded a value (shown in square brackets) that is consistent with the much more precise result for A. The value derived for A was adopted in the final analysis. The dispersion-based distance is based on a model for the interstellar electron density [29] and has an uncertainty of order 20%.

10. Future tests

In estimating the future improvements in the uncertainty of $x_{\rm B}$, and hence R and our current precision for GR tests, we need to consider that geodetic precession will lead to changes to the system geometry and hence to the aberration of the rotating pulsar beam. The effects of aberration on pulsar timing are usually not separately measurable but are absorbed into a redefinition of the Keplerian parameters. As a result, the observed projected sizes of the semi-major axes, $x^{\rm obs}_{\rm A,B}$, differ from the intrinsic sizes, $x^{\rm int}_{\rm A,B}$ by a factor $(1+\epsilon^{\rm A}_{\rm A,B})$ which depends on the orbital period, the pulse period and on the system geometry [18]. While aberration should eventually become detectable in the timing, allowing the determination of a further PK parameter, at present it leads to an undetermined deviation of $x^{\rm obs}$ from $x^{\rm int}$, where the latter is the relevant quantity for the mass ratio. The parameter $\epsilon^{\rm A}_{\rm A,B}$ scales with pulse period and is therefore expected to be two orders of magnitude smaller for A than for B. However, because of the high precision of the A timing parameters, the derived value $x^{\rm obs}_{\rm A}$ may already be significantly affected by aberration. This has (as yet) no consequences for the mass ratio $R = x^{\rm obs}_{\rm B}/x^{\rm obs}_{\rm A}$, as the uncertainty in R is dominated by the much less precise $x^{\rm obs}_{\rm B}$. We can explore the likely aberration corrections to $x^{\rm obs}_{\rm B}$ for various possible geometries. Using a range of values given by studies of the double pulsar's emission properties [49], we estimate $\epsilon^{\rm A}_{\rm A} \sim 10^{-6}$ and $\epsilon^{\rm A}_{\rm B} \sim 10^{-4}$. The contribution of aberration therefore is at least one order of magnitude smaller than our current timing precision. In the future this effect may become important, possibly limiting the usefulness of R for tests of GR. If the geometry cannot be independently determined, we could use the observed deviations of R from the value expected within GR to determine $\epsilon^{\rm A}_{\rm B}$ and hence the geometry of B.

In the near and far future, the precision of all parameters will increase further, because of the available longer time span and also the employment of better instrumentation. In a few years, we should therefore be able to measure additional PK parameters, including those which arise from a relativistic deformation of the pulsar orbit and those which find their origin in aberration effects and their interplay with geodetic precession (see [18]). On secular time scales we will even achieve a precision that will require us to consider post-Newtonian (PN) terms that go beyond the currently used description of the PK parameters. Indeed, the equations for the PK parameters given earlier are only correct to lowest PN order. However, higher-order corrections are expected to become important if timing precision is sufficiently high. While this has not been the case in the past, the double pulsar system may allow measurements of these effects in the future [24].

One such effect involves the GR prediction that, in contrast to Newtonian physics, the neutron stars' spins affect their orbital motion via spin-orbit coupling. This effect would be visible most clearly as a contribution to the observed $\dot{\omega}$ in a secular [40] and periodic fashion [50]. For the J0737−3039 system, the expected contribution is about an order of magnitude larger than for PSR B1913+16, i.e. 2×10^{-4} deg yr^{-1} (for A, assuming a geometry as determined for PSR B1913+16

[43]). As the exact value depends on the pulsars' moment of inertia, a potential measurement of this effect allows the moment of inertia of a neutron star to be determined for the first time [51]. To be successful requires the measurement of at least two other parameters to a similar accuracy as $\dot{\omega}$. While this is a tough challenge, e.g., due to the expected profile variations caused by geodetic precession, the rewards of such a measurement and its impact on the study of the equation of state of neutron stars make it worth trying.

11. Summary & Conclusions

With the measurement of five PK parameters and the unique information about the mass ratio, the PSR J0737−3039 system provides a truly unique test-bed for relativistic theories of gravity. So far, GR also passes this test with flying colours. The precision of this test and the nature of the resulting constraints go beyond what has been possible with other systems in the past. The test achieved so far is, however, only the beginning of a study of relativistic phenomena that can be investigated in great detail in this wonderful cosmic laboratory.

PK parameter	Observed	GR expectation	Ratio
$\dot{P}_b$	1.252(17)	1.24787(13)	1.003(14)
γ (ms)	0.3856(26)	0.38418(22)	1.0036(68)
s	0.99974(−39,+16)	0.99987(−48,+13)	0.99987(50)
$r(\mu s)$	6.21(33)	6.153(26)	1.009(55)

Table 2: Four independent tests of GR provided by the double pulsar as presented by Kramer et al. (2006). The second column lists the observed PK parameters obtained by fitting a DDS timing model to the data. The third column lists the values expected from general relativity given the masses determined from the intersection point of the mass ratio R and the periastron advance $\dot{\omega}$. The last column gives the ratio of the observed to expected value for each test. Uncertainties refer to the last quoted digit and were determined using Monte Carlo methods (see Supporting Online Material).

Acknowledgements

I am grateful to all my co-workers who contributed so significantly to this work. They are I.H. Stairs, R.N. Manchester, M.A. McLaughlin, A.G. Lyne, R.D. Ferdman, M. Burgay, D.R. Lorimer, A. Possenti, N. D'Amico, J.M. Sarkissian, G.B. Hobbs, J.E. Reynolds, P.C.C. Freire and F. Camilo. It is a pleasure to thank Thibault Damour and Norbert Wex for very useful discussions. The Parkes radio telescope is part of the Australia Telescope which is funded by the Commonwealth of Australia for operation as a National Facility managed by CSIRO. The National Radio Astronomy Observatory is a facility of the U.S. National Science Foundation operated under cooperative agreement by Associated Universities, Inc.

References

[1] T. Damour, G. Esposito-Farèse, Phys. Rev. D **58** (1998), 1.

[2] S. E. Thorsett, D. Chakrabarty, ApJ **512** (1999), 288.

[3] J. R. Oppenheimer, G. Volkoff, Phys. Rev. **55** (1939), 374.

[4] V. E. Zavlin, G. G. Pavlov, A&A **329** (1998), 583.

[5] G. F. Bignami, P. A. Caraveo, A. de Luca, S. Mereghetti, *SF2A-2003: Semaine de l'Astrophysique Francaise*, T. C. F. Combes, D. Barret, L. Pagani, eds. (2003), p. 381.

[6] T. H. Hankins, J. S. Kern, J. C. Weatherall, J. A. Eilek, Nature **422** (2003), 141.

[7] M. Blaskiewicz, Ph.D. thesis, Cornell University (1991).

[8] S. Johnston, D. Galloway, MNRAS **306** (1999), L50.

[9] G. Hobbs, A. G. Lyne, M. Kramer, C. E. Martin, C. Jordan, MNRAS **353** (2004), 1311.

[10] W. Zhang, F. E. Marshall, E. V. Gotthelf, J. Middleditch, Q. D. Wang, ApJ **554** (2001), L177.

[11] A. G. Lyne, R. S. Pritchard, F. G. Smith, MNRAS **265** (1993), 1003.

[12] M. Kramer, *et al.*, ApJ **593** (2003), L31.

[13] M. D. Young, R. N. Manchester, S. Johnston, Nature **400** (1999), 848.

[14] D. C. Backer, S. R. Kulkarni, C. Heiles, M. M. Davis, W. M. Goss, Nature **300** (1982), 615.

[15] J. W. T. Hessels, *et al.*, Science **311** (2006), 1901.

[16] M. A. Alpar, A. F. Cheng, M. A. Ruderman, J. Shaham, Nature **300** (1982), 728.

[17] M. Kramer, *et al.*, Science **in press** (2006).

[18] T. Damour, J. H. Taylor, Phys. Rev. D **45** (1992), 1840.

[19] T. Damour, N. Deruelle, Ann. Inst. H. Poincaré (Physique Théorique) **43** (1985), 107.

[20] T. Damour, N. Deruelle, Ann. Inst. H. Poincaré (Physique Théorique) **44** (1986), 263.

[21] M. Burgay, *et al.*, Nature **426** (2003), 531.

[22] R. N. Manchester, *et al.*, MNRAS **328** (2001), 17.

[23] V. Kalogera, *et al.*, ApJ **601** (2004), L179.

[24] A. G. Lyne, *et al.*, Science **303** (2004), 1153.

[25] M. A. McLaughlin, *et al.*, ApJ **616** (2004a), L131.

[26] M. A. McLaughlin, *et al.*, ApJ **613** (2004b), L57.

[27] `http://www.atnf.csiro.au/research/pulsar/tempo`

[28] E. M. Standish, A&A **336** (1998), 381.

[29] J. M. Cordes, T. J. W. Lazio (2002). astro-ph/0207156.

[30] J. M. Weisberg, J. H. Taylor, *Binary Radio Pulsars*, F. Rasio, I. H. Stairs, eds. (Astronomical Society of the Pacific, San Francisco, 2005), pp. 25–31.

[31] I. H. Stairs, S. E. Thorsett, J. H. Taylor, A. Wolszczan, ApJ **581** (2002), 501.

[32] J. H. Taylor, A. Wolszczan, T. Damour, J. M. Weisberg, Nature **355** (1992), 132.

[33] W. A. Coles, M. A. McLaughlin, B. J. Rickett, A. G. Lyne, N. D. R. Bhat, ApJ **623** (2005), 392.

[34] Lorimer, D. R. and Kramer, M., *Handbook of Pulsar Astronomy* (Cambridge University Press, 2005).

[35] M. Kramer, *et al.*, Annalen der Physik **15** (2006), 34.

[36] C. Lange, *et al.*, MNRAS **326** (2001), 274.

[37] S. M. Ransom, *et al.*, ApJ **609** (2004), L71.

[38] J. H. Taylor, Philos. Trans. Roy. Soc. London A **341** (1992), 117.

[39] T. Damour, R. Ruffini, Academie des Sciences Paris Comptes Rendus Ser. Scie. Math. **279** (1974), 971.

[40] B. M. Barker, R. F. O'Connell, ApJ **199** (1975), L25.

[41] M. Burgay, *et al.*, ApJ **624** (2005), L113.

[42] R. N. Manchester, *et al.*, ApJ **621** (2005), L49.

[43] M. Kramer, ApJ **509** (1998), 856.

[44] F. A. Jenet, S. M. Ransom, Nature **428** (2004), 919.

[45] I. H. Stairs, S. E. Thorsett, R. J. Dewey, M. Kramer, C. McPhee, MNRAS p. in press (2006).

[46] T. Damour, J. H. Taylor, ApJ **366** (1991), 501.

[47] K. Kuijken, G. Gilmore, MNRAS **239** (1989), 571.

[48] T. Damour, G. Esposito-Far'ese p. Submitted. (2006).

[49] M. Lyutikov, MNRAS **362** (2005), 1078.

[50] N. Wex, Class. Quantum Grav. **12** (1995), 983.

[51] T. Damour, G. Schäfer, Nuovo Cim. **101** (1988), 127.

Michael Kramer
University of Manchester
Jodrell Bank Observatory, UK
e-mail: `mkramer@jb.man.ac.uk`

Gravitation and Experiment, 101–109

Poincaré Seminar 2006

Testing Einstein in Space: The Gravity Probe B Relativity Mission

John Mester and the GP-B Collaboration

Abstract. The Gravity Probe B Relativity Mission was successfully launched on April 20, 2004 from Vandenberg Air Force Base in California, a culmination of 40 years of collaborative development at Stanford University and NASA. The goal of the GP-B experiment is to perform precision tests of two independent predictions of general relativity, the geodetic effect and frame dragging. On-orbit cryogenic operations lasted 17.3 months, exceeding requirements. Analysis of the science data is now in progress with a planned announcement of results scheduled for December 2007.

1. Introduction

Our present theory of gravity, Einstein's general relativity, is elegant, internally consistent and (so far) in agreement with observation. Yet, despite recent advances, the range of predictions tested and the precision to which experiments have tested the theory remain limited [1]. In addition, general relativity resists quantization, thwarting efforts to include the theory in a unified picture of the forces of nature. Nearly all attempts at unifying gravitation with the Standard Model result in a theory that differs form general relativity, and in particular, include additional vector or scalar couplings that potentially violate the Equivalence Principle [2,3].

Space offers the opportunity for new tests of general relativity with improved precision [4]. The use of drag compensation, first demonstrated in flight by the Discos instrument on the Triad mission [5], to reduce air drag, magnetic torque, and radiation pressure disturbances enables a uniquely quiet environment for experimentation, one not limited by seismic noise. In the following we describe a space based fundamental physics experiment, the Gravity Probe B Relativity Mission.

2. Gravity Probe B

The GP-B spacecraft was launched from Vandenberg Air Force Base on April 20, 2004 on a Boeing Delta II two-stage rocket. Although the operation plan included a 30 day orbit trim procedure, the initial orbit injection was near perfect and placed the vehicle within 100 meters of its target near the Earth's poles.

The Gravity Probe B Relativity Mission is a space based experiment developed at Stanford University with oversight by the Marshall Space Flight Center and funding from the NASA Office of Space Science. GP-B will test two predictions of general relativity, the geodetic and frame dragging effects, by measuring the precession of gyroscopes in a 642 km high orbit around the earth. The philosophy of the experiment is to reduce disturbances to levels at which general relativistic effects should be manifest without large subtraction of Newtonian drifts.

The relativistic precession of a gyroscope in a circular orbit around the earth is given by:

$$\bar{\Omega} = \left(\gamma + \frac{1}{2}\right)\frac{GM}{c^2 R3}(\bar{R} \times \bar{v}) + \left(\gamma + 1 + \frac{\alpha_1}{4}\right)\frac{GI}{2c^2 R3}\left[\frac{3\bar{R}}{R2} \cdot (\bar{\omega}_e \cdot \bar{R}) - \bar{\omega}_e\right]$$

where R is the position and v the orbital velocity of the gyroscope, I, M, and ω are the moment of inertia, mass and angular velocity of the earth, and G is the gravitational constant. For generality we include the PPN parameters γ and α_1; in general relativity $\gamma = 1$ and $\alpha_1 = 0$. The first term describes the geodetic precession, which arises from the curvature of spacetime due to the mass of the earth. General Relativity predicts that the spin direction of the gyroscope will change at the rate of 6.606 arcsec per year for a 642 km high, polar orbit. The second term, frame dragging or Lense-Thirring effect, represents the precession due to the dragging of the inertial frame by the rotation of the earth. General Relativity predicts the rate of procession of a Gravity Probe B gyroscope to be 0.039 arc sec per year (39 marcsec/yr). A polar orbit is chosen so the two precessions are orthogonal and can therefore be distinguished. Figure 1 depicts the direction of the two precessions.

3. Experimental System Overview

The small size of the relativity precessions requires that the experiment system have extreme measurement precision and that all sources of error be controlled. In order to achieve this requirement the experiment exploits the advantages of a near zero-g orbit in space and a near zero temperature in the experimental probe [6]. The experimental module consists of a helium dewar, which holds 2500 liters of superfluid helium, surrounding the experimental probe containing four gyroscopes, a quartz block, and a star tracking telescope. The dewar was designed to have an on-orbit helium lifetime of greater than 16.5 months. Actual on-orbit lifetime was 17.3 months. During the experiment, the helium is maintained at a temperature of 1.8 K by means of a porous plug venting system and the boil off

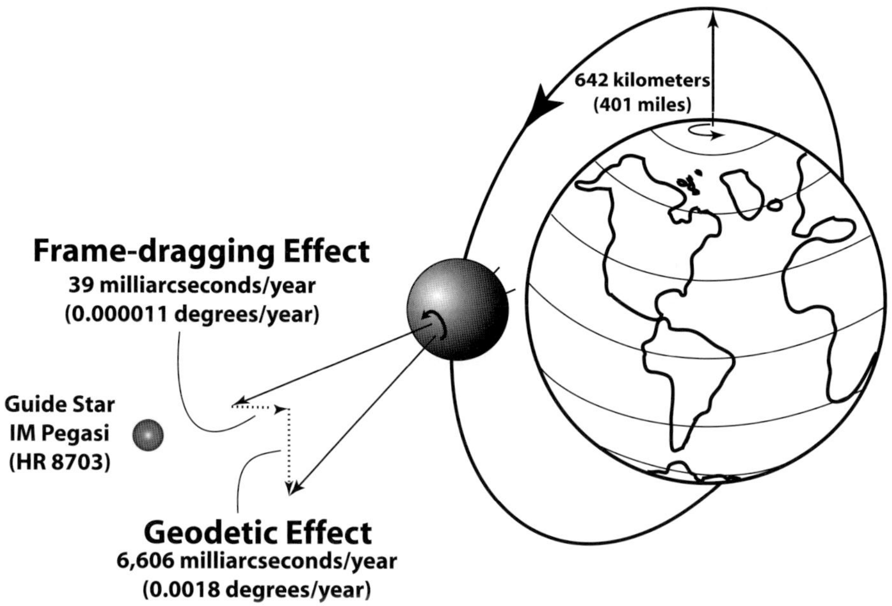

FIGURE 1. GP-B Concept

gas is used to power proportional thrusters used in drag free control. The thrusters keep the space craft centered around a gyroscope in free fall to produce residual accelerations at this gyroscope of less than $10^{-9}g$ ($g = 9.8m/s^2$). Three other gyroscopes are mounted within a rigid quartz block assembly. The quartz block provides precise positioning of the gyroscopes and the telescope, with cryogenic temperatures increasing mechanical stability. A series of windows provides an open line of sight out of the dewar. The star tracking telescope is used to point the spacecraft towards a guide star, providing a distant inertial reference with which to compare the gyro spin direction. The spacecraft rolls about the line of sight to the guide star with a period of 77.5 seconds. This averages off axis accelerations (which would contribute to Newtonian torques at the gyroscopes) to below $5 \times 10^{-12}g$ and allows the gyroscope spin direction to be measured at roll frequency, eliminating DC measurement complications.

4. Gyroscopes and Gyroscope Readout

At the heart of the gravity probe B mission are the gyroscopes. Figure 2 gives a schematic, exploded view of a gyroscope. The gyroscope is comprised of a rotor, 3.8 cm in diameter, which spins freely within the spherical cavity of a quartz housing. Newtonian torques on the gyroscope are minimized by the drag free satellite system and by controlling rotor sphericity and homogeneity. Rotors are fabricated

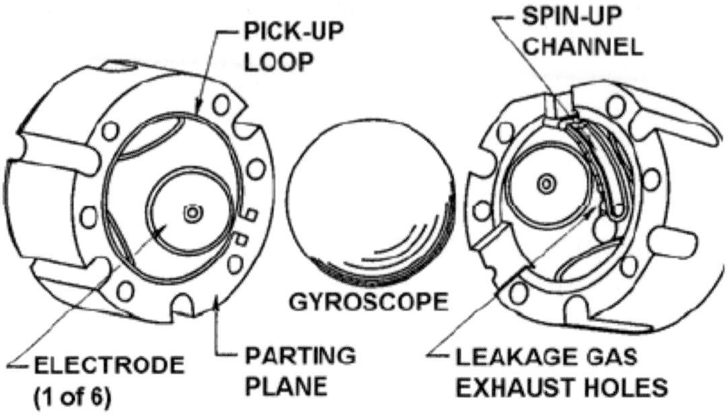

FIGURE 2. GP-B Gyroscope

from fused quartz with density inhomogeneity of less than 2 parts per million and are ground and lapped to achieve peak-to-valley asphericity of less than 25 nm. The rotors are coated with a 1.25 micron thick uniform layer of niobium which has a superconducting transition temperature of 9.2 K. The niobium coating enables the rotor to be electrostatically suspended within the housing and provides a means for sensing the gyroscope spin direction, discussed below. The housing for the rotor, shown in two halves split by a parting plane, has 3 orthogonal pairs of electrodes used to suspend and sense the position of the rotor. Once suspended, the rotors are spun-up by directing helium gas through the spin up channel. The helium gas is then pumped away to high vacuum to eliminate residual gas damping of the rotor.

The housing electrodes also provide a means of sensing rotor charge. Since the rotors are freely floating, charge can accumulate mainly due to cosmic ray, solar and South Atlantic Anomaly protons. A charge control system incorporates a UV fiber optic actuator that produces and controls photo electrons.

The gyroscope readout system must be capable of resolving changes in the rotor spin direction of less than 1 marcsec without producing interaction torques that could disturb that spin direction [7,8]. The low experimental operating temperature allows the properties of superconductivity to be exploited, both as the physical basis of the readout signal and in its detection. The readout signal is based on the magnetic field produced by the London moment of a rotating superconductor. When the superconducting niobium coated rotor is spun up, it develops a London magnetic moment aligned with its instantaneous spin axis. The London moment produces an equivalent magnetic field just outside the rotor of magnitude:

$$B_L = 1.14 \times 10^{-7} \omega_s \text{ Gauss}$$

where ω_s is the spin angular velocity. The London field is measured using a DC SQUID (superconducting quantum interference device) magnetometer. On the parting plane between the two housing halves, there is a four turn superconducting loop that couples the London moment flux to the SQUID. The gyro spin axis is aligned close to the spacecraft roll axis, which lies in the parting plane, so the London moment produces a signal modulated at roll frequency. At a spin speed at 80 hertz, the London field equals 5.7×10^{-5} Gauss. Therefore, to resolve 1 marcsec changes in spin direction a field sensitivity of 2.8×10^{-13} Gauss is required. The noise performance of Gravity Probe B SQUID readout system met this requirement, as discussed in the next section.

Such low field levels also dictate the need for extensive magnetic shielding. Ultra low DC fields of less than 10^{-7} Gauss, required to minimize flux trapping in the rotor, are produced using the expanded superconducting lead shield technique [9]. This shield, coupled with additional internal superconducting shielding surrounding each gyro, and an external cryoperm shield yield AC (roll frequency) field attenuation at the gyroscopes of greater than 2×10^{12}.

5. Telescope and Guide Star Selection

It is necessary to measure the spin direction of the gyroscopes relative to a distant reference frame, one not affected by the mass or spin of the earth. Therefore, a telescope is incorporated into the experimental module to track the position of a guide star. The star tracking telescope is of the folded Schmidt-Casagranian type with a 14.4 cm diameter aperture. It is constructed out of fused quartz and has an overall physical length of 50 cm. Two focused images are formed on the edge of roof prisms by splitting the incoming starlight with a beamsplitter. The edges of the roof prisms are perpendicular, providing two axis readout. Each prism divides the star image into two partial images whose intensities are determined using cryogenic silicon photo detectors and cryogenic preamplifiers. The relative intensities of the prism-split images determine the direction of the line of sight to the guide star. Using this signal, the spacecraft attitude is controlled to point in the direction of the guide star.

An important factor in reaching design measurement accuracy is the selection of the guide star to act as an adequate inertial reference. Uncertainties in the proper motion of the guide star propagate directly as experimental error and therefore the guide star proper motion needs to be known to high precision. Review of candidate stars led to the selection of HR8703, which is an optical star of 5.69 magnitude and is also radio star. Observations by the Harvard Smithsonian Astronomical Observatory using VLBI have establish sub 0.1 marssec/yr proper motion uncertainties [10]. VLBI observations are continuing to further reduce these uncertainties.

In parallel with space vehicle development activities, a mission operations center (MOC) was commissioned at Stanford University. The MOC conducted command and control throughout mission lifetime. The Wallops Island IONet provides communication with the TDRRS satellite communications network and ground stations at Svalbard, Norway and Poker Flat, Alaska.

6. On-Orbit Operations and Performance

On-orbit operations for GP-B lasted 17.3 months, from launch until the liquid helium cryogen was depleted. The first phase of the mission, Initial Operation and Checkout (IOC) consisted of 4.3 months of experiment setup. This was followed by 11.6 months of science data collection, and 1.4 months of post-science calibrations.

During the IOC phase many operations were required to configure the experiment for science. These included setting up the attitude and translation control system and spinning up the gyros. Initial coarse attitude was achieved using magnetic sensors and star trackers to set the spacecraft roll to a convenient operating period and to point the spacecraft to within the 60 arc-sec field of view of the telescope. Once this was realized the signal form the telescope could be included in the sensors array. Because direction to the guide star lies in the plane of the spacecraft orbit, the star is eclipsed each time the vehicle travels behind the earth. Therefore the star has to be re-acquired each orbit. After tuning of control parameters, typical drive-in times of less than 60 seconds were achieved. RMS pointing noise was less than 200 marc-sec, pointing noise at roll frequency was $\sim$ 5 marc sec.

Before gyro spin operations could proceed the gyro suspension system (GSS) electronics were activated to levitate the gyro rotors within their housings. Typical on-orbit RMS gyro position measurement noise was 0.45 nm. The control effort to keep the gyros centered provides a measure of the gyro acceleration environment. These signals were used as input to the translation control system enabling drag-free operation. Despite the loss of 2 of the 16 helium proportional thrusters, drag free performance met requirements, achieving cross track acceleration levels of $< 4 \times 10^{-12} g$ over the frequency range of 0.01 mHz to 0.1 Hz.

In order to reduce measurement noise and disturbance torques, gyro charge and trapped magnetic flux had to be minimized. The trapped flux reduction procedure consisted of heating the gyro rotors above the Nb superconducting transition temperature and then carefully cooling down within the ultra-low field environment provided by the superconducting shield system. Both the ac attenuation and dc magnetic field requirements were verified on orbit. On-orbit trapped flux levels were determined and all four gyros met requirements with the following measured values: gyro 1 – 3.0 microGauss, gyro 2 – 1.3 microGauss, gyro 3 – 0.8 microGauss, gyro 4 – 0.2 microGauss.

Rotor charge was controlled throughout the mission using a system that included the GSS sensing and UV fiber optic actuation mentioned above. After initial levitation (from which the gyros acquired charges in the range of hundreds of milli-Volts), typical charging rates throughout the mission were 0.1 mV per day. Several discharges were performed during the mission.

Gyro spin-up was accomplished by commanding a helium gas reservoir system to flow helium through the spin up channels of each gyro. In order to keep background pressures low (high pressure could interfere with gyro suspension), the gyros were spun up one at a time and helium was exhausted to space through 15 cm diameter valves. Spun gyros would slow down during the spin-up of subsequent gyros. Final gyro spin rates were in the range $\simeq$ 60 to 80 Hz, see table 1. After the final spin-up the probe residual pressure was reduced to below 10^{-11} Torr using a cryogenic bake-out procedure. This resulted in characteristic spindown period that exceeded requirements. The final procedure of the IOC phase was the ad-

	Spin rate (Hz)	Spindown period (years)
Gyro 1	79.4	15,800
Gyro 2	61.8	13,400
Gyro 3	82.1	7,000
Gyro 4	64.9	25,000

TABLE 1. Gyro spin rates and 1/e spindown periods

justment of the gyro spin axes to point within 10 arc-sec of the guide star line of site. This is required to reduce cross track torques and readout system errors. Spin axis alignment was accomplished by applying large voltages to the gyro housing electrodes in a carefully controlled sequence. This procedure also yielded an initial calibration of suspension system induced torques.

With the completion of experiment set-up, the mission entered science data phase in August, 2004. One of the main activities during the science phase was the measurement of the London moment signals from each gyro. A typical London Moment data curve is shown in figure 3. The high frequency oscillation is due to spacecraft roll, a 77.5 sec period. The broad envelope is due to guide star aberration.

The transverse velocity of the spacecraft causes a shift in the perceived guide star direction relative to the true, nearly inertially fixed location. The aberration has two components, one at the spacecraft orbit frequency and one at the annual frequency of the earth revolving around the sun. These well known amplitudes, (5.1856 arc-sec orbital, 20.4958 arc-sec annual) provide an important tool to calibrate the gyro readout/telescope readout measurement scale factor. Further scale factor calibration is provided by a commanded 60 marc-sec spacecraft dither.

After the 353 days, 6000 orbit science data phases the mission entered a post calibration stage. During this period operations were conducted to purposely increase disturbances as a means of characterizing systematic error. This included pointing the spacecraft toward other guide stars to increase cross axis gyro torques

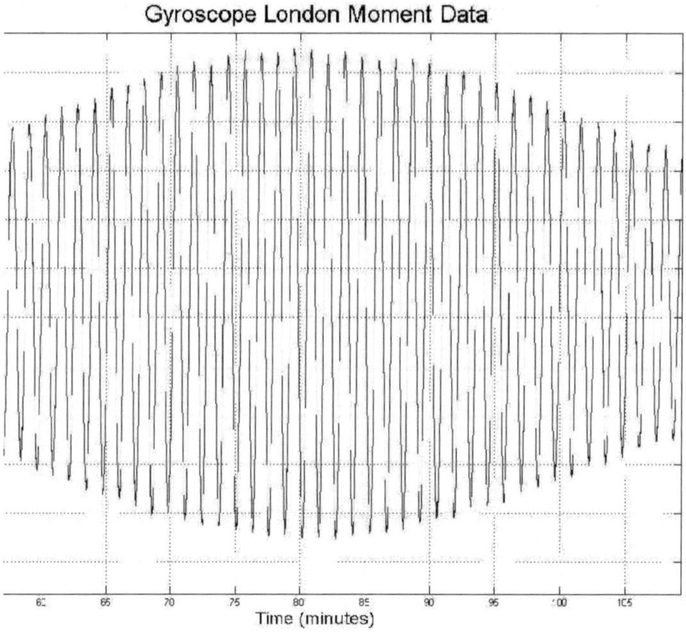

FIGURE 3. Gyro London Moment readout signal

as well as varying GSS control parameters. This phase lasted 46 days until September 29, 2005 when the last of the helium cryogen boiled off.

7. Data Analysis

With the completion of cryogenic operations the mission has entered the data analysis phase. The spacecraft communication and operations system recovered $> 99\%$ of all the data taken during the mission. These 1.5 Terabytes of data include the so called science data – the SQUID and Telescope readout signals – as well as engineering data form over 700 onboard sensors. In addition to the nominal data rate of 0.1 to 1 Hz, 200-2200 Hz "snapshot" data from key systems were also recorded and down-linked.

8. Gyro Performance

All four gyroscopes exhibited performance more than a million times better than the best inertial navigational gyros. That is, the Newtonian or disturbance drift rates were more than one million times smaller than the best navigational gyros. Indeed, the leading drift seen in the gyros, with no modeling, is that caused by

General Relativity. These unmodeled, essentially raw measurements, agree with Geodetic effect predictions to about the 1% level. To obtain a more precise Geodetic and frame dragging measurements the data analysis is focused on physical torque modeling and on refining the telescope/SQUID scale factor determination.

It is anticipated that analysis will continue into late 2007. After the analysis is completed, the separately measured proper motion of the guide star will be subtracted from the gyroscope precession rate estimates to give the final experiment result. The final results and release of data are scheduled for December 2007.

Acknowledgment

Part of this work was done during the author's stay at the Institut Henri Poincaré – Centre Emile Borel The author thanks this institution for hospitality and support.

References

[1] C.M. Will, *The confrontation between general relativity and experiment*, Astrophysics and Space Science **283**, no.4 (2003), 543–52.

[2] T. Damour, A.M. Polyakov, *The string dilaton and a least coupling principle*, Nuclear Physics B **25**, vol.B423, no.2–3 (July 1994), 532–58.

[3] I. Antoniadis, S. Dimopoulos, A. Giveon, *Little string theory at a TeV*, JHEP-Journal of High Energy Physics (2001).

[4] G.M. Keiser, *General Relativity Experiments in Space*, Adv. In Space Research **32**, no.7 (2003), 1397–1400.

[5] D.B. DeBra, *Disturbance compensation system design*, APL Technical Digest **12**, no.2 (April-June 1973), p.14–26.

[6] S. Buchman et. al., *The Gravity Probe B relativity mission*, Advances in Space Research **25**, no.6 (2000), 1177–80.

[7] J.C. Mester, J.M. Lockhart, B. Muhlfelder, D.O. Murray, M.A. Taber, *Ultralow magnetic fields and Gravity Probe B gyroscope readout*, Advances in Space Research **25**, no.6 (2000), 1185–8.

[8] J.M. Lockhart, B. Muhlfelder, G.M. Gutt, M. Luo, R.C. Clappier, T.R. McGinnis, G.R. Smith, *Optimization of a SQUID system for space*, IEEE Transactions on Applied Superconductivity **7**, no.2, pt.3, (June 1997), 2534–7.

[9] M. Taber, D. Murray, J. Lockhart, D. Frank, D. Donegan, *Production of Ultralow Magnetic Fields for Gravity Probe B*, Advances in Cryo. Eng. **39A** (1993), 161–170.

[10] S. Wang, D.-H. Gwo, K.A. Bower, L.W. Huff, J.A. Lipa, *Status of the cryogenic telescope and guide star for Gravity Probe B*, Advances in Space Research **25**, no.6 (2000), 1189–92.

John Mester and the GP-B Collaboration
Hansen Laboratory
Stanford University
Stanford CA 94305–4085, USA
e-mail: `mester@relgyro.stanford.edu`

Gravitation and Experiment, 111–138

Poincaré Seminar 2006

Instruments for Gravitational Wave Astronomy on Ground and in Space

Jean-Yves Vinet

Abstract. Gravitational Wave Astronomy progressively becomes this new window on the universe that we expected since tens of years. The technology has now reached a point where large instruments meet a level of sensitivity relevant for astrophysics. Depending on the sector of physics to be addressed, i.e. depending on the frequency domain, ground or space instruments are required. Ground based antennas are already built in Europe, in Japan and in the USA and begin to deliver science data. The ESA/NASA space mission LISA is not yet definitively approved, but a number of teams endeavour to successfully pass the coming project reviews. We review the general principles of the optical detection of gravitational waves.

1. Introduction

More and more highly relativistic objects are directly seen in the Universe and even in our Galaxy. These objects are mainly massive black holes and compact systems involving neutron stars or stellar class black holes doomed to end in an inspiral phase of variable time duration. Systems involving objects of stellar class (up to a few tens of solar masses) on compact orbits emit gravitational waves (GW) roughly in the acoustic band (from a few Hz to a few kHz), whereas more heavy systems, and especially supermassive black holes are expected to cause gravitational events of very low frequency (below 0.1 Hz). General Relativity simultaneously provides models for these processes where strong gravity deeply differ from Newtonian theory, and the right messenger (GW) for carrying the relevant information. For uncoding this information, several types of GW antennas have been proposed in the past, and a few of them survive under the form of large instruments or projects on Earth or in space. The fact that the expected dimensionless amplitudes of GW correspond to a space-time strain amplitude less than 10^{-22} is the cause of all technological issues encountered during the R&D phase of all projects. At this level, all possible environmental or instrumental noise sources must be carefully investigated, and solutions must be found in order to obtain a signal to noise ratio consistent with the observational goals. For ground based instruments like LIGO [1] and Virgo [2], the most challenging issues were the seismic insulation, the vacuum system, the mirrors processing, the laser stabilization. The ultimate residual fundamental noises are the shot noise and the thermal noise. For a space mission like LISA [3], the main issues were the long optical links, the drag-free operation and the rejection of the lasers frequency noise. This paper aims to point out and briefly discuss the physics underlying some of the technical challenges of this extreme metrology.

2. Gravitational Waves

2.1. GW emission

Gravitational waves (GW) are a consequence of Einstein's General Relativity (GR) as electromagnetic waves come from Maxwell's Electrodynamics. In the framework of Special Relativity, in a system of coordinates x^λ, an electromagnetic wave is described (in vacuum) by the vector field $A_\mu(x^\lambda)$ (4-potential) obeying the Maxwell equations. The wave propagates at velocity c, is transverse and has two polarization components. In GR, the gravitational state of spacetime is associated to its geometry through the metric tensor $g_{\mu\nu}(x^\lambda)$ obeying the Einstein equations. In the case of a gravitational wave far from its source, in a freely falling reference system, one can write:

$$g_{\mu\nu}(t,\mathbf{x}) \;=\; \eta_{\mu\nu} + h_{\mu\nu}(t,\mathbf{x}) \tag{1}$$

where $\eta_{\mu\nu} \equiv \mathrm{diag}(1,-1,-1,-1)$ is the Minkowski tensor of the locally flat background spacetime (freely falling frame), and $h_{\mu\nu}$ a very small dimensionless tensor field representing the GW amplitude. It can be shown that $h_{\mu\nu}$ can be eventually reduced to only two independent functions $h_+, h_\times$ defining the polarization state of the wave. Gravitational waves are emitted by distributions of matter/energy having a time dependent quadrupole moment. In the transverse-traceless gauge, at the first level of approximation, only the space components are significant and have an expression analogous to a retarded potential [4]:

$$h_{jk}(t,r) \;=\; \frac{2G}{c^4}\frac{1}{r}\,\partial_t^2\,\left[\Xi_{jk}(t-r/c)\right]^{\mathrm{TT}} \tag{2}$$

($r \equiv \mathbf{x}^2$) where the symbol TT refers to the projection on the transverse plane of the symmetric traceless quadrupole tensor $\Xi(t)$ defined by the volume integral:

$$\Xi_{jk}(t) \;=\; \int \rho(t,\mathbf{x})\left[x^i x^j - \frac{1}{3}\,\delta_{jk}\,\mathbf{x}^2\right] d^3x$$

where ρ is the density of matter. Further levels of approximation have been deeply investigated [5], but the preceding "quadrupole formula" gives an order of magnitude. One immediately notes the extreme weakness of the coupling coefficient G/c^4 which is the cause of all technological challenges encountered on the way to

GW astronomy. Only astrophysical events involving stars or black holes in nearly relativistic velocity regime can cause amplitudes of GW larger than 10^{-25} in the neighborhood of the Earth. The most promising candidates are final inspiralling compact binaries. The frequency domain of the waves is determined by the masses of the components of the binary. Stellar class binaries can end at 1 kHz whereas Massive Black Holes can end at a small fraction of a Hz. The observation instruments must change according to the addressed domain of frequency. Existing instruments have been designed for a sensitivity of about 10^{-23} at the middle of the bandwidth, which seemed the best feasible at the time when the preliminar R&D studies ended.

2.2. Physical signature of a GW

Being a perturbation of the geometry of spacetime, one can expect GW to produce distortions in some metrology experiments. We briefly recall the existence of narrow band solid antennas, then focus on optical experiments.

2.2.1. GW and continuous media. The first experiment proposed by Weber [6] rested on the idea that a GW could induce stresses in solids, and that on a suitably isolated solid resonator, weakly dissipative for acoustic waves, one could detect with some transducer system the resonances occurring when the GW signal overlaps its acoustical bandwidth. This idea is supported by a general relativistic extension of the linear elasticity theory [7]. A result is the modified tensor elastodynamic equation:

$$\rho \ddot{E}_{ij} - \frac{1}{2}\left[\partial_k \partial_j \Theta_{ik} + \partial_k \partial_i \Theta_{jk}\right] = -\frac{1}{2}\,\rho\,\ddot{h}_{ij}. \tag{3}$$

Where E_{ij} (resp. Θ_{ij}) is the classical strain (resp. stress) tensor, and ρ the density. If we take the origin of coordinates at the center of mass, and if we assume a GW wavelength much larger than the size of the resonator, this can be regarded as a derivative of the following vector elastodynamic equation:

$$\rho \ddot{u}^i - \partial_k\, \Theta^{ik} = -\frac{1}{2}\,\rho\,\ddot{h}^{ij} x^j \tag{4}$$

where $\mathbf{u}$ is the displacement vector. The GW amplitude appears thus as a driving internal (of tidal type) force acting on the resonator. After the controversial but negative results of Weber, several groups nevertheless built hugely improved versions of the Weber antenna. These instruments called "bar antennas" have been built in several countries [8, 9], and even larger resonators having spherical shapes are planned [10]. As any resonator, bars have a very short bandwidth (up to a few tens of Hz), so that reconstructing a waveform after a detected GW event is problematic. This is why we focus on optical experiments which are intrinsically wideband.

2.2.2. GW and light. A more direct physical effect of GW is to modulate the light distances between freely falling test masses. In vacuum, light is expected to propagate along a null geodesic, which means that the invariant element of spacetime $ds^2 \equiv g_{\mu\nu}dx^\mu dx^\nu$ is identically zero along any optical path. With the expression (1) of the metric tensor, one can suspect that the effective optical paths of photons will be perturbed.

Detectors of size much smaller than the GW wavelength. The expected frequencies of GW events for obvious reasons are much lower (at most a few kHz) than optical frequencies. In this regime, the only effect of a GW on light is to perturb the flight time of photons between two test masses (light distances). Consider a light path lying in the (x, y) plane, either along the x (north) or the y (west) axis. Consider on the other hand a GW propagating along a direction of unit vector $\mathbf{w}$,

$$\mathbf{w} = \begin{pmatrix} \sin\theta\cos\phi \\ \sin\theta\sin\phi \\ \cos\theta \end{pmatrix}. \tag{5}$$

If $h_+, h_\times$ are the two polarization components of the wave, the effect of the GW is to create a phase modulation on the two beams:

$$\Phi_{\text{north}}(t) = \frac{2\pi L}{\lambda}\left[h_+(t)(\cos^2\theta\cos^2\phi - \sin^2\phi) - h_\times(t)\cos\theta\sin 2\phi\right], \tag{6}$$

$$\Phi_{\text{west}}(t) = \frac{2\pi L}{\lambda}\left[h_+(t)(\cos^2\theta\sin^2\phi - \cos^2\phi) + h_\times(t)\cos\theta\sin 2\phi\right]. \tag{7}$$

In an interferometric configuration, where the observable is a differential phase, this gives:

$$\Delta\Phi(t) = \frac{4\pi L}{\lambda}\left[h_+(t)\,\frac{1+\cos^2\theta}{2}\,\cos 2\phi - h_\times(t)\cos\theta\sin 2\phi\right] \tag{8}$$

where it can be seen that the interferometer acts like a transducer, converting the gravitational signal into a phase and eventually into an electrical signal through some photo detector.

Detectors of size comparable to the GW wavelength. In the case of very long range optical paths (e.g. 5 Mkm in the case of LISA), one must take into account the action of the GW during light propagation. If a light beam of fixed frequency is emitted from spacecraft A and detected at spacecraft B, the nominal distance AB being L and $\mathbf{n}$ the unit vector from A to B, the physical effect detected at B is a frequency modulation. Let $\mathbf{w}$ be again the propagation unit vector of the GW, and let us define two more unit vectors mutually orthogonal in the transverse plane:

$$\vartheta = \frac{\partial \mathbf{w}}{\partial \theta}, \quad \varphi = \frac{1}{\sin\theta}\frac{\partial \mathbf{w}}{\partial \phi}$$

then the two directional functions

$$\xi_+(\theta,\phi) = (\vartheta.\mathbf{n})^2 - (\varphi.\mathbf{n})^2, \quad \xi_\times(\theta,\phi) = 2(\vartheta.\mathbf{n})(\varphi.\mathbf{n})$$

then the function

$$H(t) = h_+(t)\,\xi_+(\theta,\phi) + h_\times\,\xi_\times(\theta,\phi). \tag{9}$$

Now the observable is a relative frequency modulation, analogous to a Doppler shift [11] given by:

$$\left[\frac{\delta\nu(t)}{\nu}\right]_{A\to B} = \frac{H(t - \mathbf{w}.\mathbf{x}_B) - H(t - \mathbf{w}.\mathbf{x}_A - L)}{2(1-\mathbf{w}.\mathbf{n})} \tag{10}$$

where $\mathbf{x}_A$ and $\mathbf{x}_B$ are the positions of the two spacecraft. This is often called a "two pulses" response because a short GW pulse would have this double effect on a phasemeter at B.

3. Ground based detectors

3.1. General principles

3.1.1. Interferometers. A Michelson interferometer involves a splitter sharing a laser light into two secondary beams that recombine on it after a reflection on far mirrors, generating interferences on a photodetector. Such a device generates a phase quantum (shot) noise due to the quantum nature of light detection and characterized by the white power spectral density (PSD)

$$S(f) = \frac{2\hbar\omega}{P_L}$$

where $\lambda \equiv 2\pi c/\omega$ is the laser wavelength and P_L its power. The quantum efficiency of the photodetector has been taken equal to 1. If according to (8), the differential phase induced by the GW is at most (single and optimal polarization, normal incidence):

$$\Delta\Phi(t) = \frac{4\pi L}{\lambda}\,h(t)$$

the result is that the ultimate, shot-noise limited sensitivity of a simple Michelson having two orthogonal arms of length L is given by the white linear spectral density (LSD):

$$S_h^{1/2} = \frac{\lambda}{4\pi L}\sqrt{\frac{2\hbar\omega}{P_L}}. \tag{11}$$

It is easily seen that even with large parameters ($L = 3$ km, $P_L = 20$ W), the result ($\sim 4\,10^{-21}\mathrm{Hz}^{-1/2}$) is far from the requirements. It is seen as well that increasing these already big parameters is not so easy. Solutions have been proposed years ago by R. Drever [12].

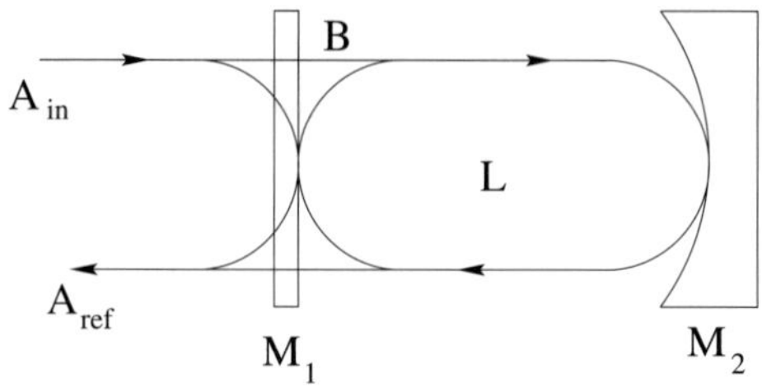

FIGURE 1. Fabry-Perot cavity

3.1.2. Resonant Fabry-Perot cavities. The leading idea is to use the properties of resonant cavities firstly for increasing the effective lengths of the arms, and secondly for increasing the effective power reaching the splitter. Consider a resonant (Fabry-Perot) cavity of length L, having an input mirror of reflectivity r_1, an end mirror of reflectivity r_2 (see Fig. 1). If a wave of amplitude $A_{\rm in}$ arrives at the input $\rm M_1$ mirror, it is partially transmitted by $\rm M_1$ and partially reflected, the transmitted part propagates to $\rm M_2$, is reflected, propagates back to $\rm M_1$ where it is partially transmitted and partially reflected. The mirrors have some weak relative losses p so that their transmission t and reflection r coefficients are related by the relative power balance $r^2 + t^2 = 1 - p$. Moreover, there must be a $\pi/2$ phase lag between the reflected part and the transmitted part, so that t and r being real numbers, we use ir as the reflection and t as the transmission operator. One can write therefore the steady state equation, assuming B as the intracavity amplitude:

$$B \;=\; t_1\, A_{\rm in} - r_1 r_2 \mathrm{e}^{2ikL}\, B \qquad (12)$$

where λ is the wavelength and $k \equiv 2\pi/\lambda$. On the other hand, the reflected amplitude is the sum of the directly reflected wave and the one partially transmitted from inside the cavity:

$$A_{\rm ref} \;=\; ir_1 A_{\rm in} + ir_2 t_1 B$$

all this put together gives the global reflectance of the cavity:

$$\mathcal{R} \;=\; A_{\rm ref}/iA_{\rm in} \;=\; \frac{r_1 + (1-p_1) r_2 \mathrm{e}^{2ikL}}{1 + r_1 r_2 \mathrm{e}^{2ikL}}.$$

Eq. (12) has the obvious solution

$$B \;=\; \frac{t_1}{1 + r_1 r_2 \mathrm{e}^{2ikL}}\, A_{\rm in}$$

which makes clear that a resonance occurs, giving a peak of stored power when the round trip phase $2kL$ is an odd multiple of π. Instead of r_1, r_2, L, a new set of relevant parameters are:

- The finesse $\mathcal{F}$ given by

$$\mathcal{F} \;=\; \frac{\pi\sqrt{r_1 r_2}}{1 - r_1 r_2}. \qquad (13)$$

- The free spectral range $\Delta\nu$ or frequency gap between two successive resonances:
$$\Delta\nu = \frac{c}{2L}. \tag{14}$$
- The linewidth $\delta\nu$ defined by:
$$\delta\nu = \Delta\nu/\mathcal{F}. \tag{15}$$

One can show that in such a cavity tuned at resonance:

- The reflectance of the cavity as a whole is:
$$\mathcal{R}_0 = 1 - \sigma$$
where the coupling coefficient σ is
$$\sigma = p\mathcal{F}/\pi \tag{16}$$
$p \equiv p_1 + p_2$ being the total relative light power loss of the cavity (thermalization, diffraction...).
- The phase change of the reflected wave on a very tiny displacement δL of the end mirror is
$$\delta\Phi = \frac{8\mathcal{F}}{\lambda}\,\delta L.$$
If we compare this to the phase change due to the same displacement without input mirror (with only the end mirror)
$$\delta\Phi = \frac{4\pi}{\lambda}\,\delta L,$$
we see that the cavity has an effect equivalent to $\mathcal{S} = 2\mathcal{F}/\pi$ round trips. For a finesse of 50, as currently planned, the result is equivalent, in terms of h, to an arm of length $L_{\text{eff}} \sim 100$ km.
- If we compute the power P_{stored} stored at resonance we obtain as long as $\sigma \ll 1$:
$$P_{\text{stored}} = \frac{2\mathcal{F}}{\pi}\,P_{\text{inc}}$$
where P_{inc} is the power reaching the input mirror. Parameter $\mathcal{S} = 2\mathcal{F}/\pi$ is called surtension coefficient. It is equal to the effective number of round trips in the cavity.

3.1.3. Recycling interferometers. These interesting properties of resonant cavities are the basis of all optical GW antennas. On Fig. 2 one can see the principles of Virgo (for instance). It can be shown that the optimal sensitivity of a shot noise limited interferometer is reached when the extinction of the outgoing beam (to photodetector) is a maximum. The whole Michelson section (the two arms plus the splitter) act therefore as a virtual mirror, and adding one more mirror (the recycler) builds a new cavity, called the recycling cavity: one may imagine that the light almost totally reflected by the Michelson re-enters the interferometer. In other words, the recycler carries out an impedance matching between the laser and the Michelson. Starting from a 20 W laser, a first resonance increases to 1 kW the power reaching the splitter. After splitting, 500 W are fed into the long cavities,

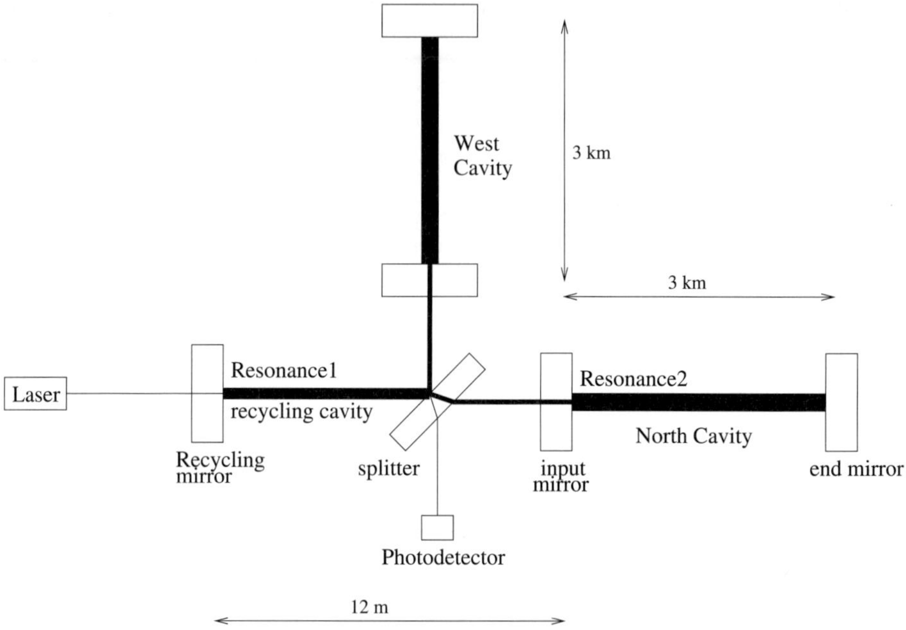

FIGURE 2. Sketch of a recycling Fabry-Perot interferometer

and the power stored in these is about 15 kW. The finesse of the arm cavities was fixed at $\mathcal{F}$ =50, increasing the effective length of the arms by a factor of about 30 giving 90 effective km, and the surtension coefficient of the recycling cavity at $\mathcal{S} = 50$. The total gain factor with respect to a simple Michelson is better than 200 according to (11), giving an LSD of h equivalent to shot noise of about $2\,10^{-23}\,\mathrm{Hz}^{-1/2}$, consistent with the requirements. This spectral density is not white any more however, because the transfer function from GW to detected phase falls to zero when the GW frequency is larger than the linewidth of the long cavities (or when the GW wavelength becomes shorter than the effective lengths of the arms as well). The complete LSD of h equivalent to shot noise is:

$$S_h^{1/2}(f) \;=\; \frac{\lambda}{8\mathcal{F}L}\sqrt{\frac{2\hbar\omega}{P_L}}\,\frac{1}{\sqrt{\mathcal{S}}}\,\sqrt{1+4(f/\delta\nu)^2} \tag{17}$$

where one sees the effect of $\mathcal{F}$ for increasing the arm length, and of $\mathcal{S}$ for increasing the laser power. Shot noise however is not the only fundamental limit to that kind of metrology.

3.2. The insulation challenge

There is no way to distinguish GW action on the space between mirrors and spurious motion of these. At the level of about $10^{-20}\mathrm{m}\cdot\mathrm{Hz}^{-1/2}$, the causes of spurious motion are a number. In particular, there is no hope to reach the fundamental limits without a very efficient insulation system.

3.2.1. Seismic insulation. The mirrors are suspended by thin wires in order to be almost free in the horizontal plane for small motions (approximation of a free fall). All optical antennas have therefore a more or less sophisticated filtering system. The complexity of the filter depends on the fixed "wall" frequency i.e. the lower frequency bound. For a wall at 50 Hz, classical acoustic filters as the LIGO's work. For pushing back the wall to 10 Hz, a more complex system was devised for Virgo. It is well known that the oscillations y of a pendulum are related to those x of its suspension point by a transfer function (TF) of the form

$$\frac{\tilde{y}(f)}{\tilde{x}(f)} = \frac{1}{1 - f^2/f_0^2}$$

where f_0 is the resonance frequency. There is thus an attenuation factor of f_0^2/f^2 for frequencies much larger than the resonance. The idea of a so-called "super-attenuator" was to construct a chain of n pendulums able to oscillate with very low resonance frequencies along all degrees of freedom (vertical and horizontal). The global TF is approximately the product of all elementary TFs, or $(f_0/f)^{2n}$ assuming comparable resonance frequencies. The pendulums are essentially heavy masses ($\sim$ 100 kg) suspended by 1m long wires giving a resonance at about 0.5 Hz for the horizontal motion, and containing steel blades whose bending stiffness has been reduced by magnets (negative spring) for the vertical motion. A global attenuation factor of about 10^{-14} at 10 Hz is obtained this way. See on Fig. 3 the details of a superattenuator.

3.2.2. Vacuum. Suppressing the refraction index fluctuations due to air pressure fluctuations requires operating in an ultravacuum. The residual hydrogen pressure must be lower than 10^{-9} mbar and 10^{-14} mbar for hydrocarbons. All the optical system and the suspensions must therefore be installed in a high quality vacuum system. The steel pipelines containing the cavities have 1.2 m diameter and are 3 km long. This represents an area of more than 20,000 m^2. An important and successful item in the R&D program was to find the thermal treatment of stainless steel able to suppress the outgassing rate at the required level in order to avoid operating with a continuous pumping. The external aspect of ground based detectors is determined by this huge vacuum system (see the example of Virgo on Fig. 4).

3.3. Fighting the thermal noise

The second fundamental limit in the sensitivity comes from the fact that the optical system essentially reads the distance between the reflecting surfaces of two mirrors (in each cavity). It is thus clear that any spurious motion of these surfaces competes with the gravitational signal. Once unessential causes of motion (sound waves, seismic vibrations...) by the insulation system are eliminated there remain sources of motion in the thermal random excitation inside all material elements

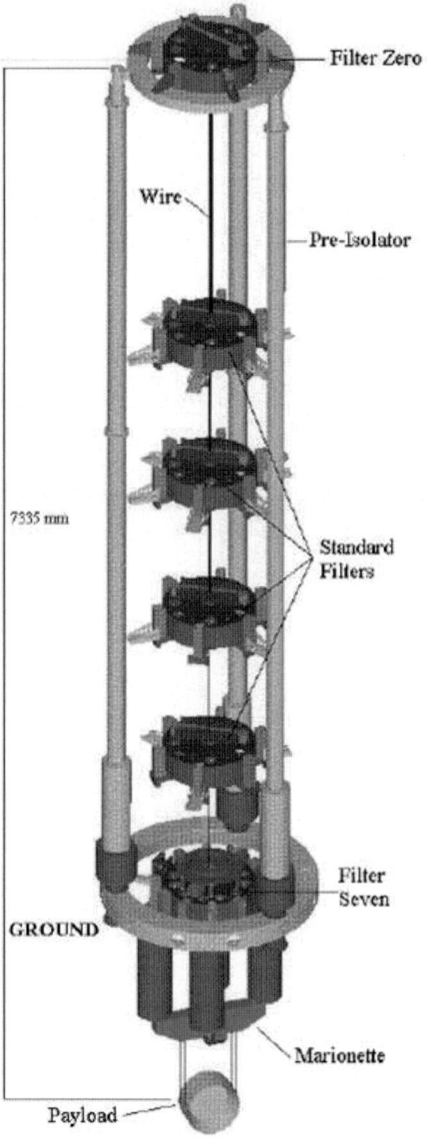

FIGURE 3. Superattenuator developed at Pisa

of the system holding the mirrors. As seen above, the filtering chain suspending the mirrors is a series of harmonic oscillators coupled with the mirror's motions. The mirrors themselves, that are thick (10 cm) and wide (35 cm diameter) silica cylinders may be considered as solid resonators and have elastodynamic modes

FIGURE 4. Global view of Virgo at Cascina (Italy)

disturbing their shapes and resulting in an apparent displacement. The resulting noise in the readout system is called thermal noise. Three different sectors of thermal noise can be distinguished:

- The thermal excitation of the pendulum chain suspending the mirror holder. The corresponding spectral density fixes the wall frequency at 10 Hz
- The thermal excitation of the wires holding the mirrors from the last pendulum stage. The resonances of these wires are called "violin modes". These resonances can be sharpened by using high Q materials and weldings. The present trend is to use monolithic silica suspensions.
- The internal modes of the mirrors substrates give a PSD of noise with a low frequency tail dominating all noises in the 100 Hz region.

The region of 100 Hz being very interesting from the astrophysics point of view, a number of efforts have been spent for finding ways of reducing the mirrors thermal noise. This is why we put a special emphasis on this particular sector of past and current R&D efforts. Let us discuss the proposed ideas.

3.3.1. Cooling. The PSD of thermal noise is proportional to the temperature T, so that cooling is an obvious good idea. But the LSD is proportional to $\sqrt{T}$ so that in order to gain 1 order of magnitude in sensitivity, one must reach cryogenic temperatures. Some R&D has been carried out by a Japanese team [13] in the LCGT project.

3.3.2. New materials. The PSD of thermal noise of any oscillator depends also on its mechanical quality factor. The quality factor of a compound system is determined by the intrinsic mechanical dissipation rate in the used materials but also by the way they are assembled. In the present situation, for instance in Virgo, the mirrors are suspended by thin steel wires. More specifically it has been shown [14] that the low frequency tail of the PSD of displacement equivalent to thermal noise is:

$$S_x(f) = \frac{4k_BT}{\pi f}\,\phi U \tag{18}$$

where k_B is the Boltzmann constant, ϕ a loss angle (inverse of a quality factor) and U the strain energy stored in the solid substrate under a pressure distribution having the profile of the readout optical beam and normalized to 1 N resulting force. The second parameter to play with is the loss angle ϕ. It seems that it is difficult to have loss angles less than 10^{-6} with synthetic silica. This is why it has been proposed to use sapphire instead. Unfortunately sapphire has bad optical properties, so that it could not be used for transmitting mirrors.

3.3.3. Alternative Beam geometries. The third parameter on which to act is U, which leads to look for beam profiles that decrease that (virtual) mirror strain. In the current situation, the optical beams circulating in the interferometers are the Gaussian beams emitted by standard lasers in which optical power is focused on a small spot at the center of the mirror. A way of calculating U has been found in [15] when total axisymmetry is assumed. A first approximation, valid if the hot spot radius is small compared to the mirror size, and a Gaussian TEM_{00} mode of parameter w, having thus an intensity profile of the form:

$$I(r) = \exp\left[-2r^2/w^2\right]$$

is simply, regarding the mirror substrate as an semi-infinite medium:

$$U = \frac{1-\sigma^2}{2\sqrt{\pi}Yw}$$

where Y is the Young modulus of the substrate and σ its Poisson ratio. This makes clear that it is desirable to increase parameter w. For larger values of w, the assumption of a semi-infinite medium cannot be kept, and the dimensions of the (cylindrical) substrate, its radius a and its thickness h must enter the model. For summarizing the result, U is the sum of two contributions:

$$U = U_0 + \Delta U$$

that can be computed separately. Some notation must be recalled. The $J_\nu(x)$ are the ordinary Bessel functions and $\{\zeta_k,\ k \in \mathbb{N}^*\}$ the discrete family of all non-zero solutions of $J_1(\zeta) = 0$. Let us note $x_k \equiv \zeta_k h/a$ and $q_k \equiv \exp(-2x_k)$. If now the intensity distribution in the readout beam is $I(r)$, we can define its Fourier-Bessel coefficients as:

$$p_k = \frac{2\pi}{J_0^2(\zeta_k)} \int_0^a I(r) J_0(\zeta_k r/a)\, r\, dr$$

and an auxiliary parameter ξ as

$$\xi = \sum_{k>0} \frac{p_k J_0(\zeta_k)}{\zeta_k^2}. \tag{19}$$

Finally, U_0 can be expressed as the following series:

$$U_0 = \frac{1-\sigma^2}{\pi a Y} \sum_{k>0} \frac{J_0^2(\zeta_k) p_k^2}{\zeta_k} \frac{1 - q_k^2 + 4q_k x_k}{(1-q_k)^2 - 4q_k x_k^2} \tag{20}$$

and ΔU, using (19) as:

$$\Delta U = \frac{a^2}{6\pi h^3 Y} \left[\left(\frac{h}{a}\right)^4 + 12\sigma\xi \left(\frac{h}{a}\right)^2 + 72(1-\sigma)\xi^2 \right]. \tag{21}$$

At this point, all the thermal noise PSD amounts to compute the beam-profile coefficients p_k. For a TEM_{00} mode as above:

$$p_{k,0}^{(0)} = \frac{1}{J_0^2(\zeta_k)} \exp\left[-\frac{\zeta_k w^2}{8a^2}\right].$$

A way of obtaining a more homogeneously distributed light power by using special mirrors such that the cavity eigenmodes are "flat top" beams has been proposed [16]. If we adopt a simple model in which the intensity is assumed constant on a disk of radius $b < a$ and zero outside, the corresponding profile coefficients p_k are:

$$p_{k,flat} = \frac{2aJ_1(\zeta_k b/a)}{b\zeta_k J_0^2(\zeta_k)}$$

For instance, with $b = 11.3$ cm, for a Virgo mirror ($a = 17.5$ cm, $h = 10$ cm) a gain factor of about 3 could be achieved in the LSD with respect to the present situation [17] . This kind of optical modes are however obtained in a Fabry-Perot cavity by using non spherical mirrors. Some numerical and experimental R&D studies have been carried out [18] to test the operation of such cavity a from the point of view of optical stability under small misalignments.

Recently it has been proposed [19] to use high order TEM modes, to obtain a more homogeneous power distribution. These modes at the same time allow a better noise reduction and keep a spherical wavefront. The profile coefficients for a Laguerre-Gauss mode $\mathrm{LG}_{n,m}$ are:

$$p_{k,m}^{(n)} = \frac{1}{J_0^2(\zeta_k)} \exp\left[-\frac{\zeta_k w^2}{8a^2}\right] L_m\left(\frac{\zeta_k w^2}{8a^2}\right) L_{m+n}\left(\frac{\zeta_k w^2}{8a^2}\right)$$

where the $L_n(x)$ are the Laguerre polynomials. For instance a $\mathrm{LG}_5^{(5)}$ mode with $w = 3.5$ cm used as a readout beam for a mirror of diameter $2a = 35$ cm and thickness $h = 10$ cm (Virgo parameters) could achieve a gain of about 5 in sensitivity with respect to the present situation in Virgo, without significant increase in diffraction losses. An important point is that this allows to keep spherical mirrors.

3.4. Issues in Optical technology

3.4.1. Technology. The shot-noise limited sensitivity of $2 \cdot 10^{-23}\,\mathrm{Hz}^{-1/2}$ that has been shown above to be theoretically feasible relies on a good reflectance of the Michelson subsystem, i.e. a reflectance allowing to get the required surtension $\mathcal{S} \sim 50$ in the recycling cavity. It is easily seen that the maximum $\mathcal{S}$ is:

$$\mathcal{S}_{\max} = \frac{1 - p_r}{1 - (1 - p_r)(1 - p_s)^2(1 - \sigma)^2}$$

where p_r, p_s are the losses at the recycling mirror and at the splitter respectively, whereas σ is the coupling rate of the cavities defined above (16). These losses are dominated by the coupling rate (by a factor comparable to the finesse), so that a rough estimate of the maximum gain is:

$$\sqrt{\mathcal{S}_{\max}} = \frac{1}{\sqrt{2\sigma}} = \frac{1}{\sqrt{4p\mathcal{F}/\pi}}$$

For having $\mathcal{S} \geq 50$ with $\mathcal{F} = 50$, the overall losses must therefore be less than 300 ppm. These losses involve not only thermalization of light, but also scattering (roughness of the reflecting surfaces), diffraction (aberration) and misalignments, so that this figure of 300 ppm is demanding. Scattering losses scale as $1/\lambda^2$. A special synthetic silica has been developed specially for Virgo & LIGO in order to make very low absorption substrates, a specific polishing protocole has been defined with a manufacturer for obtaining superpolished surfaces. Then the polished surface becomes reflecting after a coating process in which stacks of dielectric layers with alternative low and high refractive indices are deposited. The wavelength of $\lambda \sim 1.064\,\mu$m is the best found compromise allowing powerful light sources (Nd:YAG lasers) and low scattering losses (< 1 ppm). This coating process was first developed for very small highly reflecting mirrors involved for instance in laser gyros. Extension to large surfaces (~ 35 cm) was a challenge. A specific facility has been built at the IPN-Lyon, able to process large diameter samples in a clean environment [20].

3.4.2. Simulation. It was and still is difficult to assess the optical requirements without a numerical modelization of the interferometer. This is why a special code has been developed [21]. The main point is to represent light propagation from a given plane to a next one at a finite distance L. In the paraxial approximation of wave optics, if we call $A(x, y)$ the wave complex amplitude at abscissa $z = 0$, coordinates (x, y) being defined in the transverse plane and $B(x, y)$ the diffracted amplitude at $z = L$ is obtained from the Fresnel integral, provided the diffraction angles are not too wide (paraxial):

$$B(x, y) = \int_{\mathbb{R}^2} K_L(x - x', y - y')\, A(x', y')\, dx'\, dy' \qquad (22)$$

where the function $K_L(x,y)$ is the diffraction kernel (λ is the wavelength and $k \equiv 2\pi/\lambda$):

$$K_L(x,y) = -\frac{i}{\lambda L} \exp\left[ik\frac{x^2+y^2}{2L}\right].$$

(22) being a convolution product, it can be expressed under the form of a Fourier transform. If the Fourier coordinates that conjugate from (x,y) are denoted by (p,q), we have:

$$\tilde{B}(p,q) = \tilde{K}_L(p,q) \times \tilde{A}(p,q)$$

The Fourier transform $\tilde{K}_L$ (propagator) has an explicit expression:

$$\tilde{K}_L(p,q) = \exp\left[-iL\frac{p^2+q^2}{2k}\right].$$

The point is that reduced to Fourier transforms, the propagation problem can be treated via Fast Fourier Transform algorithms which allows very efficient codes. The way of propagating a wave amplitude is thus

- take the 2D-FFT of $A(x,y)$
- multiply by the propagator (it has been computed once for all)
- take the reciprocal 2D-FFT and get $B(x,y)$.

This method is especially efficient in the case of resonant cavities. For instance, consider the equation (12) for intracavity amplitude $A_{\rm ic}(x,y)$ inside a cavity from an incoming amplitude $A_{\rm in}(x,y)$. It can be generalized as:

$$A_{\rm ic} = M_{1,\rm trans}A_{\rm in} + M_{1,\rm ref}.\mathcal{P}.M_{2,\rm ref}.\mathcal{P}.A_{\rm ic} \quad (23)$$

where the linear operator P refers to the sequence detailed above and the $M_{\rm i,ref,trans}$ to phase plates equivalent to the mirrors properties (geometry of the surface, transmission chart). An explicit algebraic solution formally exists but provides no realistic computation scheme. A realistic method is to take an initial guess for $A_{\rm ic}$ (for instance the theoretical mode assuming perfect mirrors) and iterating (23) until a given accuracy is met. This is the basic principle of DarkF, a code developed within the Virgo collaboration after [21]. This method allows to treat imperfect beams and imperfect mirrors. It is possible to import in the code measured maps for all mirrors and give tilt angles, detunings in order to check the performances of the resulting virtual interferometer versus the nominal estimations.

3.5. Planned spectral sensitivity

Owing to the preceding discussion, the overall spectral sensitivity is an envelope, resulting at low frequency ($<$ 50 Hz) from the pendulum thermal noise, in the intermediate range around 100 Hz from the mirror's internal thermal noise, and in the upper part of the spectrum from shot noise. At the resonances of the suspension wires, thin peaks appear. The foreseen sensitivity curve, after reduction of non-essential noises has the shape summarized on Fig. 5.

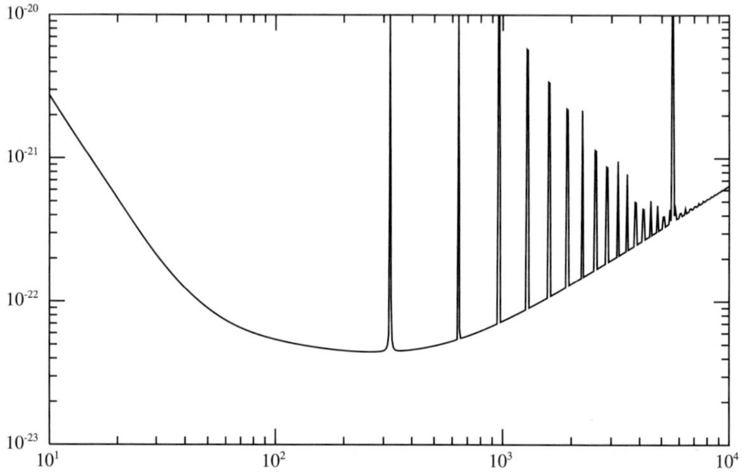

FIGURE 5. Nominal (theoretical) spectral sensitivity of Virgo

3.6. Frequency Stabilization

Frequency fluctuations of the source laser result in a specific noise on the detection channel. If the random process $\delta\nu(t)$ denotes these frequency fluctuations, the corresponding phase fluctuations on the dark fringe are

$$\delta\Phi(t) = \frac{d}{L}\frac{2\pi L}{c}\,\delta\nu(t) \tag{24}$$

where L is the mean length of the arms and d their difference. For a simple short Michelson interferometer, it is easy to carefully tune the arm lengths to obtain an arbitrarily low level of noise. For a long baseline interferometer with resonant cavities, the effective length $L_{\text{eff}} = (2\mathcal{F}/\pi) \times L$ depends not only on the geometrical length L of the arms, but also on the finesses of the cavities. These finesses depend in turn on the reflection coefficients of the mirrors, so that the noise level is eventually determined by the ability of technology to produce mirrors with very close reflectivities, allowing to make as symmetrical cavities as possible. We have seen that the phase shot noise LSD is

$$S_\Phi^{1/2}(f) = \sqrt{\frac{2\hbar\omega}{\mathcal{S}P_L}}.$$

We can rewrite (24) in terms of LSD, by asking the frequency fluctuations to produce a phase noise lower than the shot noise:

$$\frac{S_{\delta\nu}^{1/2}}{\nu_L} < \frac{\lambda}{4\mathcal{F}L}\frac{\mathcal{F}}{\Delta\mathcal{F}}\sqrt{\frac{2\hbar\omega}{\mathcal{S}P_L}}$$

where $\Delta\mathcal{F}$ represents the difference between the finesses of the North and West cavities. With the same values as above, assuming $\mathcal{S}P_L \sim$1 kW reaching the splitter, a mean cavity finesse of $\mathcal{S} \sim 50$, this is:

$$S_{\delta\nu}^{1/2} < \frac{10^{-8}}{\Delta\mathcal{F}/\mathcal{F}}\ \mathrm{Hz}\cdot\mathrm{Hz}^{-1/2}$$

with a symmetry rate of 1% for the finesse, we obtain a requirement of

$$S_{\delta\nu}^{1/2} < 10^{-6}\ \mathrm{Hz}\cdot\mathrm{Hz}^{-1/2}.$$

This very strong requirement is satisfied by at least two stages of frequency control. A first stage is a servo loop using error signals by comparison of the laser frequency with a reference passive, very stable resonant cavity (long term stabilization). A second stage is a servo loop on the common mode of the two long arms (short term stabilization). The result is one the most stable oscillators in the present metrology status. Obtaining at the same time an output power of about 20W is obtained via the injection technique, in which the stabilized laser light enters a powerful slave laser whose mode is locked on the master wave.

3.7. Data Analysis

The order of magnitude of the sensitivity likely makes a very poor signal to noise ratio at least in the present generation of antennas. This is why special signal processing techniques have been developed for extracting GW signature from the dominant instrumental noise background. Expected signals are:

- short bursts (a few ms) possibly produced by supernova or exotic cosmic string events
- permanent waves emitted by fast pulsars having some quadrupolar moment
- chirps emitted during the inspiral/merging/ringdown process of binary coalescence

For the detection of binary black hole (BBH) coalescence, the common strategy of all groups is to use the matched filtering technique. It is possible to accurately describe the inspiral phase either by using the Parametrized Post Newtonian approach [5] or the Effective One Body method [22], or by numerical simulations. The result provides families of templates (each corresponding to a point in the parameter space). An example can be seen on Fig. 6 a BBH with 30 solar masses each. The signal is assumed to enter the detection band at the date at which its frequency is 20 Hz. The total duration is less than 2s. Detection amounts to look for a correlation peak between the interferometer output and a bank of templates.

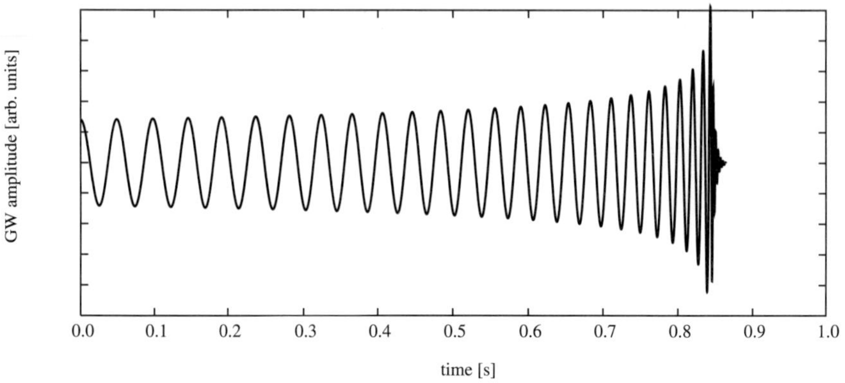

FIGURE 6. Coalescence signal for two black holes of equal mass (30 solar masses)

3.8. Present status

The two american LIGO instruments at Hanford (WA) and Livingston (LA) are already operating at their nominal sensitivity. The size of the LIGO antennas is 4 km. At the Hanford site, one more antenna is installed in the same vacuum pipe, with 2 km size. The French-Italian Virgo at Cascina (Italy) of size 3 km comes at the end of its commissioning phase. The Virgo sensitivity is already comparable to the american antennas at high frequency (see Fig. 7). A German-British antenna GEO600 of size 600 m is operating since 2002 near Hannover. Its sensitivity remains less than larger antennas (see Fig. 8), but alternative optical designs could allow to reach a comparable sensitivity at the price of a reduced bandwidth. A Japanese antenna of size 300 m, TAMA [24] is operating since 2003, the best sensitivity being about $10^{-21}\mathrm{Hz}^{-1/2}$ (see Fig. 9).

4. The LISA mission

LISA is the present status of a very old idea initiated in the seventies and aiming to receive and analyze very low frequency GW from sources involving massive black holes. There is on Earth a "wall" at a few Hz that forbids, due to direct Newtonian attraction of test mirrors by ground motions, going to lower frequencies. The solution is therefore in space. LISA is an ambitious ESA/NASA joint mission which consists of three spacecraft forming a triangle of 5 Mkm a side, in orbit around the Sun 50 Mkm behind the Earth. The three spacecraft are optically linked by six Nd:YAG laser beams. The GW signature is read on the six Doppler data flows (beat note of the incoming light against the local oscillator). LISA is expected to fly in 2014 if the project passes a review in 2008 against a few

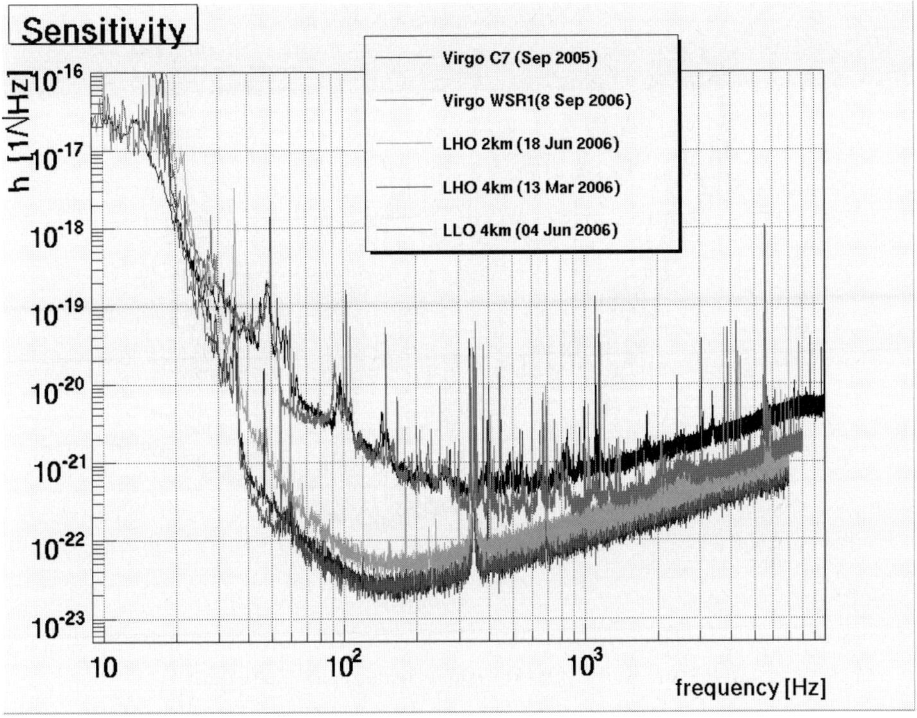

FIGURE 7. Compared sensitivities of LIGO and Virgo antennas (LLO=LIGO Livingstone Observatory, LHO=LIGO Hanford Observatory, WSR=Weekly Science Run, C=Commissioning)

other fundamental physics missions, and if the technological demonstrator LISA-Pathfinder is successful.

4.1. Orbital configuration

The stability of a large triangular formation on heliocentric orbits is not trivial. It can be shown (e.g. [25]) that it is possible by combining slightly elliptical, slightly inclined orbits. More specifically, let L be the inter-spacecraft distance ($L \sim 5\,10^9$m) and R the radius ($R \sim 1.5\,10^{11}$m) of the almost circular terrestrial orbit. Let us define the small parameter $\alpha = L/2R \sim 1/60$. We can choose simultaneously for the orbit of spacecraft #1 an inclination angle of ϵ with respect to ecliptic, and an eccentricity of e. The right choice is:

$$\epsilon \;=\; \arctan\left[\frac{\alpha}{1+\alpha/\sqrt{3}}\right] \quad \text{and} \quad e \;=\; \sqrt{1+\frac{2\alpha}{\sqrt{3}}+\frac{4\alpha^2}{3}}-1.$$

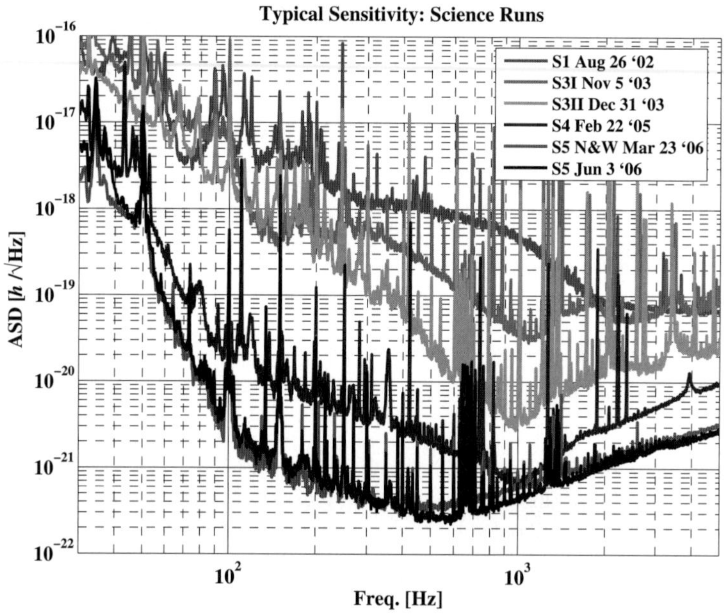

FIGURE 8. GEO600 typical sensitivity (after[23])

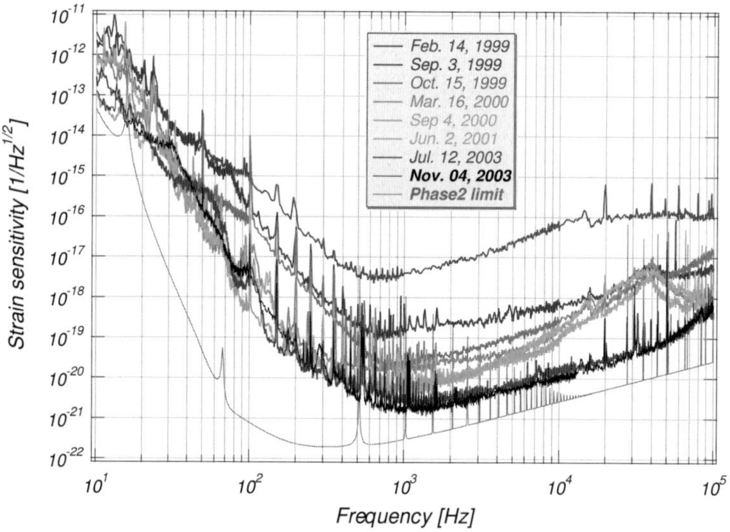

FIGURE 9. TAMA300 typical sensitivity (after[24])

In barycentric coordinates (centered on the Sun with (x, y) axes in the ecliptic and fixed with respect to far stars), the motion of spacecraft #1 has the parametric form:

$$\left\{\begin{array}{lcl} x &=& R(\cos E_1 - e)\cos\epsilon \\ y &=& R\sqrt{1-e^2}\,\sin E_1 \\ z &=& -R(\cos E_1 - e)\sin\epsilon \end{array}\right. \tag{25}$$

where $E_1(t)$ is the so-called eccentric anomaly implicitly defined by

$$E_1 - e\,\sin E_1 \;=\; \Omega t$$

where $\Omega \equiv 2\pi/(1\text{ year})$. The orbits of spacecraft #2 and 3 are obtained by

- Shifting by 120 degrees the eccentric anomaly, so that

$$E_i - e\,\sin E_i \;=\; \Omega t - (i-1)\frac{2\pi}{3} \quad (i = 1, 2, 3)$$

- Rotating the semi-major axes by 120 degrees in the (x, y) plane, so that the motions of all spacecraft are parametrized by:

$$\left\{\begin{array}{lcl} X_i &=& x_i\cos\theta_i - y_i\sin\theta_i \\ Y_i &=& x_i\sin\theta_i + y_i\cos\theta_i \\ Z_i &=& z_i \end{array}\right. \tag{26}$$

  where $\theta_i \equiv (i-1)\times 2\pi/3$, and where the (x_i, y_i, z_i), $i = 1, 2, 3$ are parametrized according (25) with the E_i.

The result is that the three spacecraft are located in a plane making an angle of 60 degrees with respect to ecliptic, with mutual distances constant at first order in α, making a triangle rotating around its mass center with a 1 year period (see Fig. 10). "At first order in α" means that a more accurate evaluation shows a deformation of the triangle, and inter-spacecraft distances variable by about 100,000 km. It is possible to reduce this "flexing" to less than 50,000 km by slightly increasing the 60 degrees angle [26].

4.2. Drag free operation

At the level of $\Delta L/L \sim 10^{-22}$, meaning a $\Delta L \sim 5\,10^{-12}$m, it is clear that perturbations caused by solar winds must be strongly rejected. It is therefore planned to operate LISA under the drag free regime. This means that the spacecraft protecting shell is served on an internal reference mass by a capacitor readout system. The free falling reference mass plus the readout system form an accelerometer. This kind of accelerometer has been imagined and successfully flown by the French ONERA on several space missions [27], but here the targetted readout noise is about $3\,10^{-15}\,\mathrm{m}\cdot\mathrm{s}^{-2}\mathrm{Hz}^{-1/2}$ in a frequency range from 10^{-4} Hz to 10^{-1} Hz. The model relevant for LISA has been developed following analogous principles. Controlling the spacecraft position with respect to the test mass requires controlled forces. These forces are applied by micro-thrusters. Two systems are being proposed and will be tested in the "LISA Pathfinder" demonstration mission.

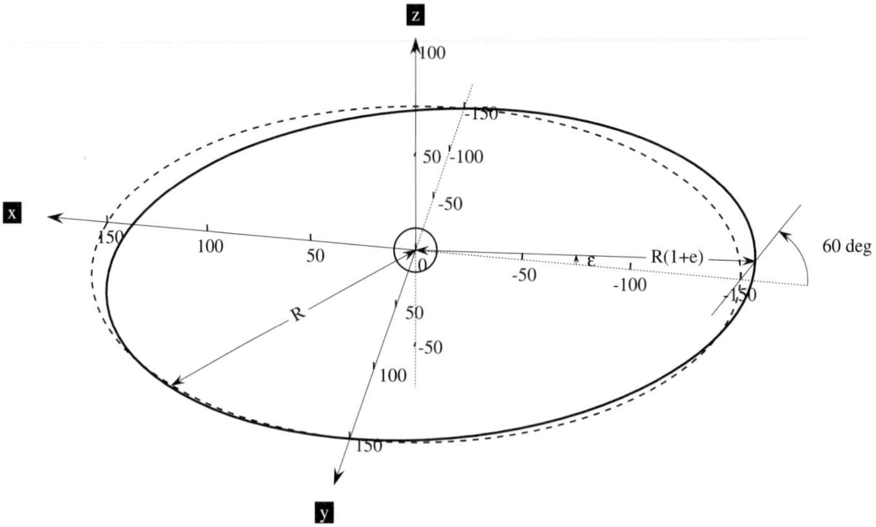

FIGURE 10. Solid: Orbit of spacecraft #1. Dashed: Earth's orbit (ecliptic)

4.3. Data flow

In a very simplified scheme, the LISA readout system involves six phasemeters, each delivering its own data flow. If we call $C_i(t)$ $(i = 1, 2, 3)$ the instantaneous frequencies aboard the three spacecraft , the apparent Doppler measurement on board spacecraft #1 for light coming from spacecraft #2 is (counterclockwise) according to (10):

$$V_1 = \left[\frac{\delta\nu(t)}{\nu}\right]_{2\to 1} = \frac{H(t - \mathbf{w}.\mathbf{x}_1) - H(t - \mathbf{w}.\mathbf{x}_2 - L_3)}{2(1 - \mathbf{w}.\mathbf{n}_3)} + C_1(t) - C_2(t - L_3) + s_1(t) \tag{27}$$

(see Fig. 11) $s_1(t)$ accounts for the shot noise generated by the detection process on board spacecraft #1. We take into account the fact that the triangle may be not equilateral, so that we have to deal with three different lengths L_i, $i = 1, 2, 3$. The data flows V_2, V_3 can be obtained by cyclic permutation of the indices. For the clockwise links, we get

$$U_2 = \left[\frac{\delta\nu(t)}{\nu}\right]_{1\to 2} = \frac{H(t - \mathbf{w}.\mathbf{x}_2 - L_3) - H(t - \mathbf{w}.\mathbf{x}_1)}{2(1 + \mathbf{w}.\mathbf{n}_3)} + C_1(t - L_3) - C_2(t) + s_2(t). \tag{28}$$

The data flows U_3, U_1 are obtained by cyclic permutation of indices.

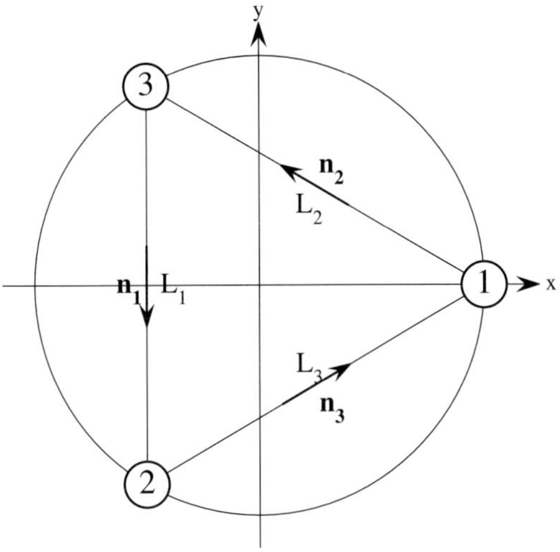

FIGURE 11. Notations for the LISA geometry

It seems that the signal ($\sim 10^{-22}$), due to the huge asymmetry between the optical path of the long link (5 Mkm) and the local path ($\sim$ 1 m) is largely dominated by the frequency noises of the lasers (the C_i, $\sim 10^{-13}$ Hz$^{-1/2}$ under closed stabilization loop). Fortunately, the number of data flows allows some redundancy leading to a dramatic reduction of the noise.

4.4. Time Delay Interferometry

We can define three delay operators D_i via their action on any function of time f:

$$(D_i f)(t) \;=\; f(t - L_i).$$

If we consider the part of the Doppler data due to laser noise, we can write:

$$\begin{aligned} U_1 &= D_2C_3 - C_1 \\ U_2 &= D_3C_1 - C_2 \\ U_3 &= D_1C_2 - C_3 \\ V_1 &= C_1 - D_3C_2 \\ V_2 &= C_2 - D_1C_3 \\ V_3 &= C_3 - D_2C_1. \end{aligned} \tag{29}$$

It has been remarked that some combinations of the U_i, V_i give an identically zero result, and can be regarded as noise canceling. Use of such noise canceling combinations was proposed by M. Tinto [28] and called "Time Delay Interferometry" (TDI). The algebraic structure of TDI has been found and explained in [11]. The simplest example is found by considering the C_i, U_i, and the V_i as vectors $\mathbf{C}$, $\mathbf{U}$, $\mathbf{V}$ and the delay operators D_i as the components of a vector operator $\mathbf{D}$. The sum $\mathbf{U}+\mathbf{V}$ has the algebraic signature of a curl:

$$\mathbf{U}+\mathbf{V} \;=\; \mathbf{D}\times\mathbf{C}.$$

It is now clear that the "divergence" of $\mathbf{U}+\mathbf{V}$ is identically zero, so that

$$\mathbf{D}\cdot(\mathbf{U}+\mathbf{V}) \;=0 \;\Rightarrow\; \sum_{i=1}^{3} D_i U_i + \sum_{i=1}^{3} D_i V_i \;=\; 0.$$

Each noise canceling ("silent") combination y can thus be represented by a 6-tuple $Y=(p_i,q_i)$ of polynomials in the formal variables D_i, acting on the data 6-tuple $U=(V_i,U_i)$:

$$y \;=\; \langle Y|U\rangle \;=\; \sum_{i=1}^{3}(p_i V_i + q_i U_i).$$

The basis of TDI is the set $\mathcal{S}$ of all silent Y. It has been shown [11] that $\mathcal{S}$ has the algebraic structure of a first module of syzygies on the ring of formal polynomials. This means that any element of $\mathcal{S}$ can be obtained by a linear combination whose coefficients are polynomials in D_i, of generators of $\mathcal{S}$. A generating part of $\mathcal{S}$ has been found by [28], containing:

$$\zeta \;=\; (\mathbf{p},\mathbf{q}) \;=\; (D_1,D_2,D_3,D_1,D_2,D_3)$$

with the new silent 6-tuple:

$$\alpha \;=\; (1,\; D_3,\; D_1D_3,\; 1,\; D_1D_2,\; D_2)$$

plus its two successive circular permutations (of indices and of locations in the sub-3-tuples):

$$\beta \;=\; (D_1D_2,\; 1,\; D_1,\; D_3,\; 1,\; D_2D_3)$$

$$\gamma \;=\; (D_2,\; D_2D_3,\; 1,\; D_1D_3,\; D_1,\; 1).$$

Any combination of α,β,γ applied to the data 6-tuple $\mathbf{U}$ is thus laser noise free. It can be shown that the same combination is still sensitive to GW. Generator ζ (often called "symmetric Sagnac") strongly attenuates the GW signals at low frequency. There is no hope however to suppress the shot noise nor the noise coming from the accelerometer readout system, because those are purely local noises (not transmitted to other spacecraft with some delay). The global sensitivity curve for a typical TDI combination ("Michelson") is shown on Fig. 12. It assumes one year integration time for a permanent source, with a signal to residual noise ratio of 5, and an average on the angular coordinates of the source. In reality, the situation is more complex because there are six lasers, not three, and the propagation times between two spacecraft are not reciprocal, due for instance to the Sagnac effect in rotating frames, and are even variable in time due to the flexing effect. But the preceding method remains valid in principle, up to improvements [29, 30].

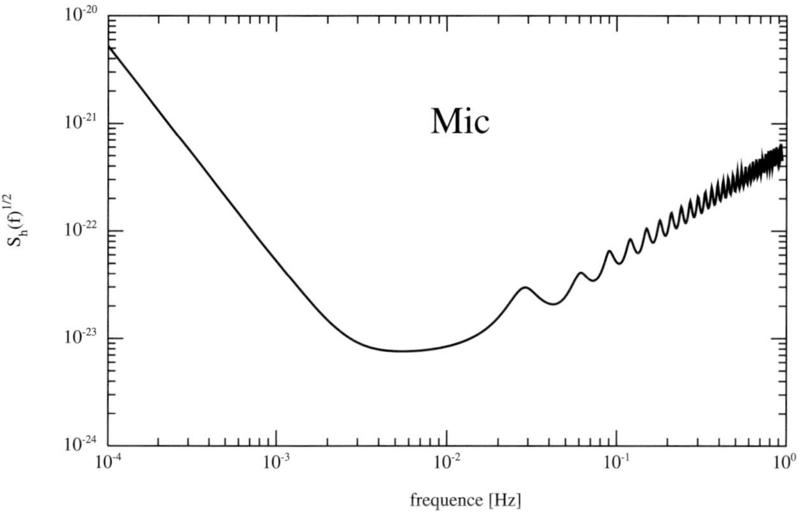

FIGURE 12. Mean spectral sensitivity of LISA for the "Michelson" TDI combination

4.5. Data analysis

Owing to the low frequency domain of sensitivity, the sampling frequency may be taken at a fraction of a Hz, so that the flow of data down to Earth is consistent with the bandwidth of the microwave link. A preprocessing by TDI generators is necessary, which implies a good knowledge of the instantaneous inter-spacecraft distances. Depending on the kind of source to be studied, several strategies may be developed.

A major point is the existence of a foreground of GW noise generated by the population of galactic compact binaries (involving neutron stars, white dwarfs, black holes) whose orbital frequencies (times 2) fall within the detection band of LISA. The GW amplitude resulting from all these monochromatic sources is analogous to a stochastic background. These objects are a huge number (e.g. $\sim 10^8$ white dwarf binaries, see [31]), All of these produce a so-called confusion noise the spectral density of which is dominant at very low frequency (from 0.1 to 1 mHz). For detecting a particular binary, it is possible to find optimal combinations of the generators [32] and even combinations giving a zero result, allowing to selectively suppress known sources [33, 32] for a "coronographic" operation of LISA.

The signals generated by or around black holes are of two types.

- Stellar class objects orbiting supermassive black holes have complex inspiralling trajectories ending by a capture. The models for such events are called

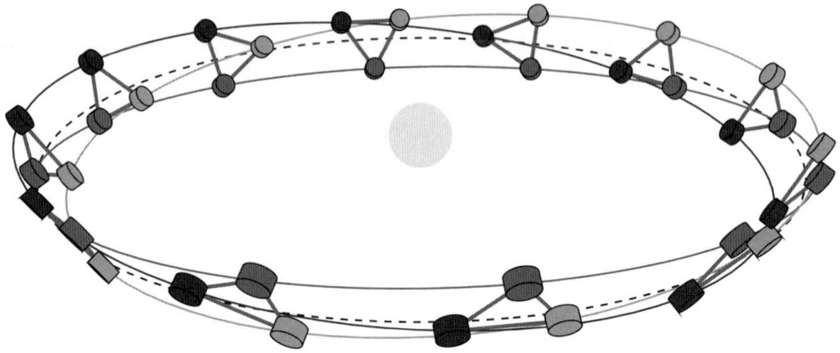

FIGURE 13. Orbital evolution of LISA after "LISACode"

Extreme Mass Ratio Inspirals (EMRI). The GW emitted during EMRI have a complex frequency structure [34]. These models depend on a number of parameters making difficult a matched filtering approach. Moreover, situations may happen in which several bodies are involved resulting in a perturbed process escaping the "simple" model. Time-Frequency methods based on wavelet transforms seem a relevant tool.

- Supermassive Binary Black Holes (SMBBH) are expected to inspiral on a long time period. The final phase could be observed during several years of LISA operation. A matched filtering approach is possible, and Time-Frequency methods as well.

4.6. Simulators

The closest date of launch of the mission being 2014, the various algorithms building the Data Analysis System must be developed and tested with synthetic data. Two data simulators have been coded in the United States and one in France:

- Synthetic LISA at the Jet Propulsion Laboratory (Pasadena, California) [35]
- The LISA Simulator at Montana U. [36]
- LISACode by LISA-France (APC, Observatoire de la Côte d'Azur) [37] (see Fig. 13)

The simulators compute the orbital motion (26) of each spacecraft and the corresponding transfer function (27,28) for the GW signals. The GW amplitudes for several kind of sources and angular locations are read from files and the result is given in terms of TDI generators.

5. Conclusion

The first generation of ground GW antennas is now beginning to deliver science data. Technological improvements will probably be needed to achieve a better sensitivity which could improve the rate of detected events. Ideas already exist for improving the laser power and for reducing the thermal noise, which were the main obstacles to overcome. The fate of the large space antenna LISA will depend on the willing of the Relativistic Astrophysics community to continue the international cooperation that began so long ago.

References

[1] `http://www.ligo.caltech.edu`

[2] `http://wwwcascina.virgo.infn.it`

[3] `http://lisa.jpl.gov`

[4] Kip S. Thorne, in 300 years of gravitation, Edited by S.W.Hawking and W. Israel, Cambridge U.P. 1987.

[5] L. Blanchet, Phys. Rev. D **72** (2005), 044024.

[6] Joseph Weber, in Gravitation and Relativity, Edited by Chiu and Hoffmann, Benjamin 1964.

[7] J.-Y. Vinet, Ann. Inst. Henri Poincaré, Vol. XXX n3 (1979) p. 251.

[8] L. Baggio et al., Phys. Lett. **95** (2005), p. 081103.

[9] P. Astone et al., Class. Quant. Grav. **23** (2006), S57–S62.

[10] A. de Waard et al., Proceedings of the 6th Amaldi Conference on Gravitational Wave, Okinawa, Japan (2005), (to be published in Jour. of Phys. : Conference series).

[11] S.V. Dhurandhar, R.K. Nayak and J-Y. Vinet, Phys. Rev. D **65** (2002), 102002.

[12] Ronald Drever, in *Gravitational Radiation*, edited by N. Deruelle and T. Piran (North Holland 1983) p. 321.

[13] T. Uchiyama et al., Class. Quantum Grav. **21** (2004), S1161–S1172.

[14] Yu. Levin, Phys. Rev. D **57** (1998), p. 659.

[15] F. Bondu, P. Hello and J.-Y. Vinet, Phys. Lett. A **246** (1998), p. 227–236.

[16] E. D'Ambrosio, Phys. Rev. D **67** (2003), 102004.

[17] J.-Y. Vinet, Class. Quantum Grav. **22** (2005), p. 1395.

[18] J. Agresti et al., LIGO technical documents, `http://www.ligo.caltech.edu`

[19] B. Mours, E. Tournefier, J.-Y. Vinet, Class. and Quantum Grav. **23** (2006), p. 5777–5784.

[20] `http://lma.in2p3.fr`

[21] P. Hello, N. Man, A. Brillet, J.-Y. Vinet, Journal de Physique I vol **2** (1992), p. 1287–1303.

[22] A. Buonanno, T. Damour, Phys. Rev. D **59** (1999), 084006.

[23] `http://www.geo600.uni-hannover.de`

[24] `http://tamago.mtk.nao.ac.jp`

[25] S. Dhurandhar, R. Nayak, S. Koshti, J.-Y. Vinet, Class. Quantum Grav. **22**, 3 (2005), p. 481–487.

[26] R. Nayak, S. Koshti, S. Dhurandhar, J.-Y. Vinet, Class. Quantum Grav. **26** (2006), p. 1763–1778.

[27] P. Touboul, A. Bernard, C.R. Acad. Sci. Paris Sér. IV (2001) p. 1271.

[28] M. Tinto, J. Armstrong, F. Estabrook, The Astroph. Journal **527** (1999), p. 814–826.

[29] M. Tinto, F. B. Estabrook, J. W. Armstrong, Phys. Rev D **69** (2004), 082001.

[30] R. Nayak, J.-Y. Vinet, Phys. Rev. D **70** (2004), p. 102003.

[31] G. Nelemans, L. Yungelson, S. Portegies Zwart, A&A **375** (2001), 890–898.

[32] R. Nayak, S. Dhurandhar, A. Pai, J.-Y. Vinet, Phys. Rev. D **68** (2003), 122001.

[33] M. Tinto, S. Larson, Class. Quantum Grav. **22** 10 (2005), S531–S535.

[34] S. Hughes, Class. Quantum Grav. **18** (2001), p. 4067–4073.

[35] `http://www.vallis.org/syntheticlisa/`

[36] `http://www.physics.montana.edu/lisa/`

[37] `http://www.apc.univ-paris7.fr/SPIP/article.php3?id_article=164`

Jean-Yves Vinet
ARTEMIS
Observatoire de la Côte d'Azur
B.P. 4229
06304 Nice Cedex 4 - France
e-mail: `Jean-Yves.VINET@obs-nice.fr`